TITANIUM
Metallic Pollutants in the Aquatic Environment

Rachel Ann Hauser-Davis
Oswaldo Cruz Foundation, Fiocruz
Rio de Janeiro, Brazil

CRC Press
Taylor & Francis Group
Boca Raton London New York

CRC Press is an imprint of the
Taylor & Francis Group, an **informa** business

A SCIENCE PUBLISHERS BOOK

First edition published 2024
by CRC Press
2385 NW Executive Center Drive, Suite 320, Boca Raton FL 33431

and by CRC Press
4 Park Square, Milton Park, Abingdon, Oxon, OX14 4RN

Library of Congress Cataloging-in-Publication Data (applied for)

ISBN: 978-1-032-33344-1 (hbk)
ISBN: 978-1-032-33345-8 (pbk)
ISBN: 978-1-003-31924-5 (ebk)

DOI: 10.1201/9781003319245

Typeset in Palatino Linotype
by Prime Publishing Services

Preface

The use of metals throughout human history has been paramount to societal advances since the first discovery of copper in 9000 BC. Metals have played an essential role in the improvement of basic activities such as transport and agriculture, as well as in tools and weaponry, progressing to arts and craft, construction, packaging, infrastructure, medical applications, and energy production throughout time. The so-called "metals of antiquity" comprise the seven metals which humans found use for in prehistoric times and on which civilization was based on, namely gold, silver, copper, tin, lead, iron, and mercury. These metals played a significant role in the development of early civilizations and had a profound impact on the economic, cultural, and technological development of early civilizations. This list has increased significantly, and today consists in about 94 metals, depending on the applied definition (RSC 2022). Today's society is completely dependent on these compounds. Many of them, however, are still understudied in several contexts, especially regarding their toxicity to living organisms.

Metal contaminants reach aquatic environments from various sources, such as industrial activities, agricultural practices, urban runoff, wastewater discharge, and accidental spills resulting in harmful environmental effects, such as biodiversity losses leading to deleterious cascade events, including aquatic food web alterations and overall ecosystem degradation, as well as negative human health effects due human exposure to contaminants through the consumption of contaminated seafood. In this scenario, this series on metallic pollutants aims to presents an integrated and holistic discussion on several understudied metals, including toxicity aspects, physiological biota effects and health risk assessments to both aquatic organisms and humans.

This first book is on titanium.

Contents

The Discovery of Titanium

The metal titanium (Ti) was first discovered mixed with a grey-black sand containing white fine grains in 1791 by the clergyman, chemist, and mineralogist Reverend William Gregor in his parish, named Menachan, in Cornwall, Great Britain (Weeks 1932). When analysing the sand, he determined two metal oxides, an iron oxide attracted to a magnet, and 45.25% of an unidentified white metallic oxide, which he named menachanite in honour of the parish in which it was discovered (Weeks 1932). However, although he reported his discovery in both German and French science journals, Crell's Annalen and Observations et Mémoires sur la Physique, respectively (Gregor 1791a,b), these papers did not cause much of a stir at the time, and menachanite was, unfortunately, soon forgotten (Weeks 1932).

Four years later, in 1795, the Prussian chemist and first professor of chemistry at the University of Berlin, Martin Heinrich Klaproth, rediscovered titanium when analysing a rutile specimen originated from Boinik (in Hungary at the time) given to him as a gift by Count Wurben of Vienna (Jagnaux 1891). After a thorough investigation of this metal's properties, he rechristened it menachanite titanerde, or titania, named after the Titans of Greek mythology (Klaproth 1801), and confirmed that Gregor's previously described sample did indeed contain Ti.

Pure Ti, however, defied isolation for several years, as it is usually found in impure form, i.e., as an oxide containing nitride and/or cyanide, and the chemistry techniques of the time did not allow for adequate purification (Weeks 1932). It was only several hundred years later, in 1910, that Mathew Arnold Hunter, an American chemist and metallurgist, isolated Ti at 99.9% purity by heating pure Ti chloride and Ti sodium samples at 700 to 800°C under pressure in an airtight steel cylinder (Hunter 1910), which shall be described in further detail in the following sections. Following this milestone, Ti properties have made it an important metal in today's society.

Titanium Properties

Titanium, with an atomic number of 22 and atomic weight of 47.867, is a silvery grey transition metal belonging to the Group IVb of the periodic table, with an electronic configuration of $[Ar]3d^24s^2$. This metal presents the highest strength-to-weight ratio among all metals (Donachie 1988), with a hardness of 6 on the Mohs scale (Britannica 2022). The physicochemical properties of this element are depicted in Table 1 (RSC 2023).

Pure titanium is non-magnetic, only exhibiting paramagnetic behavior, i.e., displaying weak magnetic properties when exposed to a magnetic field due to the arrangement of its electrons and the absence of unpaired electrons in its atomic structure, with a magnetic moment of about 0.01 Bohr magnetons (Livraghi et al. 2011). However, Ti's magnetic properties can be influenced by impurities or alloying with other elements, where it can then can exhibit weak ferromagnetic or antiferromagnetic properties (Esther Rubavathi et al. 2019).

Titanium contains five non-radioactive isotopes (Ti^{46}, Ti^{47}, Ti^{48}, Ti^{49}, and Ti^{50}) and eight unstable isotopes (Table 2). The main and most stable oxidation state is 4+, while states 3+ and 2+, although less stable and typically less prevalent and are also found in nature (Britannica 2022). While the chemistry of Ti in the +2 state is highly restricted, many compounds are formed by titanium in the +3 state, such as trichloride ($TiCl_3$), a crystalline form widely employed as a catalyst in the production of polypropylene, and in the 4+ state, where Ti forms compounds such as titanium dioxide (TiO_2), titanium tetrachloride ($TiCl_4$), and titanium tetrafluoride (TiF_4),

Table 1. Physicochemical titanium properties.

Property	Description
Strength-to-weight ratio	High (40%)
Atomic number	22
Atomic weight	47.88
Density at 25°C	4.5 g cm^{-3}
Boiling point	3287°C
Melting point	1668°C
Tensile strength	220 Mpa
Shear modlue	42.0 GPA
Hardness	70
Elongation at breaking point	54%
Poisson ratio	0.34
Thermal expansion coefficient from 20°C to 100°C	8.90 μ m °C
Thermal conductivity	17 W mk^{-1}
Electrode potential	0.20 V
Ionic radius	0.680 A
Electronegativity	1.54
Coefficient of linear expansion at 25°C	8.5×10^{-6} K^{-1}
Latent heat of fusion	20.9b kJ mol^{-1}
Latent heat of sublimtion	464.7 J mol^{-1}
Latent heat of vaporization	397.8 kJ mol^{-1}
Specific heat capacity at 25°C	0.523 J g^{-1} K^{-1}

Table 2. Titanium isotope characteristics.

Isotope	Atomic mass (Da)	Isotopic abundance (%)
^{46}Ti	45.952	8.25
^{47}Ti	46.951	7.44
^{48}Ti	47.947	73.72
^{49}Ti	48.947	5.41
^{50}Ti	49.944	5.18

among others, with the most important comprising Ti oxide (TiO_2), with hundreds of different applications (Britannica 2022).

In its pure form Ti is a lustrous white metal. Due to a passive, highly stable adherent oxide coating that protects this metal from oxidation, primarily composed of titanium dioxide (TiO_2), Ti is slow to react with water and air at ambient temperatures, making it highly corrosion-resistant and not very water-soluble (Ishii et al. 2003). This passive oxide coating is even able to regenerate through a process called re-passivation (Bocchetta et al. 2021). In most cases, the oxide layer is typically a few nanometres thick but can be further enhanced through various surface treatments, such as anodizing, heat treatment or chemical passivation (Szota et al. 2020).

Titanium also presents a similar strength to steel with only half the density, and is about 5-fold stronger than aluminium, also presenting high resistance to extreme temperatures and pressure (Qiu and Guo 2022), as well as biocompatibility (Raines 2010). These properties make titanium a versatile material used in a wide range of applications, and its unique combination of strength, durability, corrosion resistance, and biocompatibility makes it highly valued in various industries, including in extreme environments such as space and the deep-sea (Guo et al. 2020), which shall be discussed in the following sections.

Titanium Sources and Extraction Processes

Titanium is the 9th most abundant element on Earth, constituting about 0.63% of the earth's crust, ranking behind only oxygen, silicon, aluminum, iron, calcium, sodium, potassium, and magnesium in terms of abundance. Titanium makes up about 0.63% of the Earth's crust by weight, and, although it is not as abundant as some other elements, it is still considered a relatively common element in the Earth's crust (Britannica 2022). Titanium oxide, on the other hand, is present in minerals from 15 to over 95%, found in most rocks, and soils, including sandy and clayey soils, and occurring mostly in alluvial and volcanic formations (Seagle 2019). Although Ti is one of the most abundant elements in the Earth's crust, it is not always easily extractable from its ores, mainly as it is usually bonded to other elements. This has led to the development of specialized and complex processing techniques, such as smelting, roasting, and chemical separation, to obtain pure titanium dioxide for various industrial applications, which, in turn, increase Ti costs compared to other more common metals, such as iron or aluminium (Faller and Froes 2001).

This metal has been identified in about 140 minerals to date, with the most common depicted in Table 3 (Seagle 2019).

In nature, TiO_2 mainly occurs in three structural states, namely rutile, anatase and brookite (Meinhold 2010). In addition to these primary mineral sources, titanium dioxide can also occur in other forms, such as brookite and perovskite. However, these forms are less common and not as economically significant for titanium extraction.

Table 3. Main titanium-containing minerals, their chemical formulas and common impurities.

Mineral	Chemical formula	Common impurities
Anatase	TiO_2 (polymorph)	Al, Ca, Cr, Fe, Mn, P, Th, V, Y, alkali metals, alkaline earth metals, rare earth metals
Brookite	TiO_2 (orthorhombic variant)	Fe, Nb, Ta
Ilmenite	$FeTiO_3$	Mg, Mn, Fe
Perovskite	$CaTiO_3$	Fe, Na, Nb, Ni, Th, U, rare earth metals
Rutile	TiO_2 (polymorph)	Fe, Nb, Ta
Titanite (formerly sphene)	$CaTiSiO_5$	Al, Fe

Rutile is found as a constituent of various types of rocks, including igneous rocks, metamorphic rocks, and sedimentary rocks. It owes its name a its deep red colour observed in some rutile specimens in transmitted light (derived from the Latin *rutilus*), although yellowish and brownish rutile colours are also common (Clarks and Williams-Jones 2004). Rutile contains about 95 % of Ti in the form of titanium dioxide (TiO_2), with Ti occurring in the Ti^{4+} oxidation state, and ilmenite, which contains from 50 to 65% of TiO_2 (Meinhold 2010, Seagle 2019). This mineral is typically crushed, purified, and chemically treated to produce pure titanium.

It is noteworthy that ilmenite, another mineral, can also be processed into rutile. Ilmenite, a metallic black or dark brown mineral whose name is derived from the Ilmen Mountains in Russia, where it was first discovered (Brunet and Hauviller 1977), is commonly found in igneous rocks and sedimentary deposits and is often associated with other minerals, such as rutile and zircon. Most of the world's Ti reserves are mostly found as illmenite, with major ilmenite deposit regions found in Australia South Africa, America, India, and Brazil, although about one fifth of Australia's reserves are found in the form of rutile, also found in India and South Africa, Serra Leone, Canada, China and India, albeit at much lower amounts (Statista 2022).

Global Ti resources in the form of mineral deposits total about 700 thousand metric tons of TiO_2 present in illmenite and

49 thousand tons of TiO_2 found in rutile, distributed throughout several countries, such as China, Australia, the United States (primarily in Florida and Virginia), India (mainly concentrated in the coastal states of Odisha, Tamil Nadu, and Kerala), South Africa, Ukraine (in the Crimean Peninsula and the Zhytomyr region), Russia, Norway, Malaysia, Sierra Leone, and Canada, among others (Seagle 2019). Chinese stores account for about one-third of all global reserves (mainly in the coastal provinces of Hainan, Guangdong, and Guangxi) (Wu and Zhang 2006), found entirely as ilmenite, and totalling about 230 million metric tons of titanium dioxide content, ranking first in 2021 (Statista 2022), followed by Australia (major deposits found in Western Australia, Queensland, and New South Wales), India, Brazil, Norway, Canada, South Africa (major deposits in the Eastern Cape and KwaZulu-Natal provinces), Mozambique, Madagascar, Ukraine, and the United States (Statista 2022).

As mentioned previously, pure metallic Ti was first obtained in 1910 by Mathew Arnold Hunter at the Rensselaer Polytechnic Institute, in a batch process now known as the Hunter Process (Hunter 1910). Small amounts of pure Ti were also produced in 1925 by Dutch chemists Anton Eduard van Arkel and Jan Hendrik de Boer when they discovered the iodide process, through a reaction with iodine followed by decomposition of the formed vapours over a hot filament (Van Arkel and de Boer 1925), now termed the van Arkel–de Boer process. Titanium applications remained mainly laboratorial until 1932, when William Justin Kroll, a Luxembourgish metallurgist, produced it by reducing titanium tetrachloride ($TiCl_4$) with calcium, refining this process with magnesium and sodium in the following years, finally achieving the Kroll process (Greenwood and Earnshaw 1997). Briefly, this extraction process comprises a first step that converts TiO_2 present in ores into $TiCl_4$, which is then reduced to Ti using magnesium or sodium. The ore is then heated with chlorine and coke at about 1,000°C, producing $TiCl_4$, which is then reduced using either magnesium or sodium under an argon atmosphere, also at about 1,000°C. When cooled, the reaction mixture is crushed, and mixed with dilute hydrochloric to react with any excess magnesium to form more magnesium chloride. All the magnesium chloride is then dissolved in the water present in the mixture and the remaining Ti is processed further for further purification (Sohn 2020).

Other Ti extraction routes have been developed over the years (for a detailed review, see Takeda et al. 2020). Briefly, according to Takeda et al. (2020), Borchers and Hupperts obtained Ti by electrolysis in molten alkaline-earth halide as far back as 1904 (Takeda et al. 2020). The final product was, however, probably heavily contaminated by oxygen and carbon. Other modified electrolytic methods were developed in the 1960s, based on titanium chloride electrolysis in molten chloride. However, industrialization failed for the most part, due to low current efficiency caused by the formation of multi-valent titanium ions, difficult separation between titanium deposits and the electrolytic baths, low productivity due to slower reaction rate than in metallothermic reductions, and low space electrolytic cell utilization efficiency. The 1980s saw a renewed interest in Ti extraction, employing both electrolytic and metallothermic reductions using diverse raw materials and reducing agents in different combinations. Several processes displayed high potential, such as the FFC Cambridge process, an electrochemical method developed by George Chen, Derek Fray, and Thomas Farthing between 1996 and 1997 at the University of Cambridge, lowering costs and obtaining higher Ti yields (Krebs 2006). The FFC process was further modified in the 2000s to produce low-cost titanium powder, although the Kroll process is still the predominant commercial Ti production method today (Krebs 2006). Figure 1 depicts the most common titanium mineral processing methods.

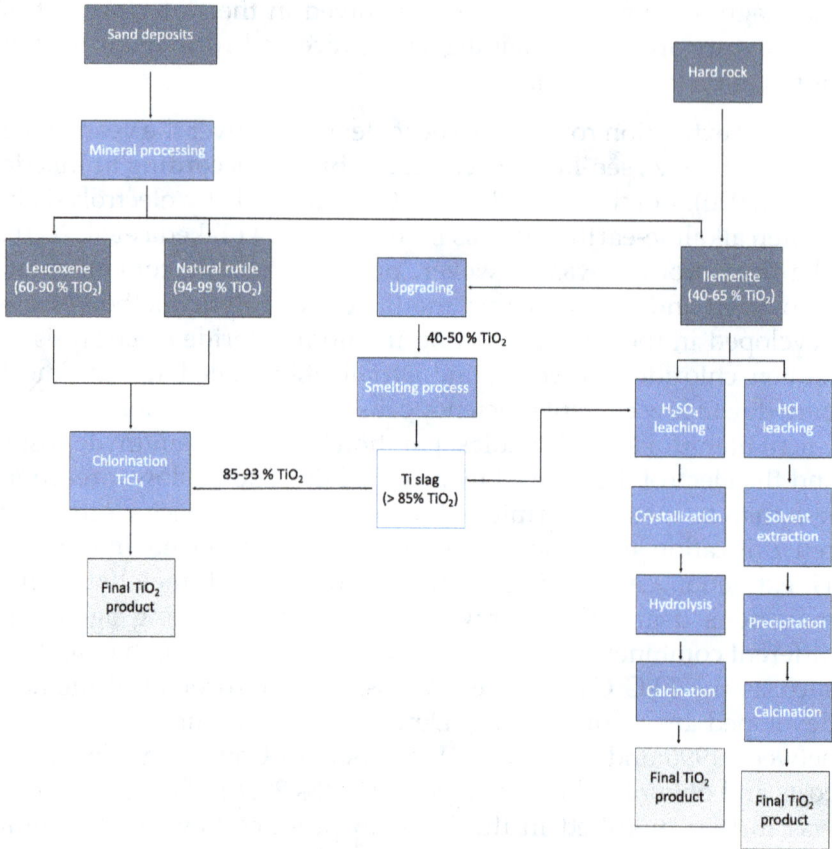

Figure 1. The most common titanium mineral processing methods (adapted from Laksmanan et al. 2014).

Titanium Applications

Due to its unique properties, including the aforementioned corrosion resistance, low density, high tensile strength to density ratio and thermal strength, Ti is now widely applied in the most diverse fields, with industrial, agricultural, scientific, environmental, medical, space and technological applications. This meyal is also a part of a recent categorization termed Technology Critical Elements (TCEs) (Watari et al. 2020), comprising a group of minerals and elements that play a crucial role in various high-tech industries and emerging technologies, which includes, besides Ti, Rare Earth Elements (REEs), Platinum Group Metals (PGMs, including platinum, palladium, and rhodium) lithium, cobalt, graphite, indium, tantalum, gallium, and antimony, among others (Romero-Freire et al. 2019). This classification considers the increasing uses of certain elements that were previously minimally exploited and that "form a strong industrial base, producing a broad range of goods and applications used in everyday life and modern technologies", most likely on a global scale (Cobelo-Garcia et al. 2015), leading to increased attention focusing on their potential environmental impacts (EU 2020a, Batley and Campbell 2022). The demand for TCEs has increased significantly with the rapid growth of technology-driven industries. However, their limited availability and geopolitical considerations surrounding their production and supply have raised concerns about their long-term availability and sustainability (Nuss and Blengini 2018). Efforts are being made to diversify sources, improve recycling and recovery technologies, and develop alternative materials to reduce reliance on TCEs and ensure the continued development of advanced technologies (Krishna et al. 2023).

One of the main Ti properties that makes it employed in hundreds of applications worldwide is its hardness, with a rating of around 70 on the Rockwell hardness scale (HRB) (Britannica 2022). It is usually alloyed with other elements and not used alone, due to limited solid solution strengthening capabilities. When alloyed, its hardness is significantly increased under these conditions, forming form titanium alloys that exhibit excellent strength and hardness. When alloyed, Ti is used in very high amounts, of up to 60% (Bai 2015), as a high tensile strength to density ratio is required to increase strength.

The microstructures of materials play a crucial role in determining their mechanical properties. In this regard, titanium alloys are classified into α, near-α, $\alpha + \beta$, metastable β, or stable β, based on their microstructure at room temperature, with the α and β phases being the most common (Dai and Song 2019). These alloys employ α-stabilizers, such as Al, O, N, C, β-stabilizers, such as Mo, V, Nb, Ta, Fe, W, Cr, Si, Co, Mn, among others, and neutrals, such as Zr, respectively (Dai and Song 2019). Dai and Song indicate that both α and near-α titanium alloys exhibit excellent corrosion resistance but exhibit limited strength at low temperature, while, $\alpha + \beta$ alloys offer higher strength due to the presence of both α and β phases (Liu et al. 2004).

The most common titanium alloy is Ti-6Al-4V (90% titanium, 6% aluminum, 4% vanadium), which accounts for more than 50% of the titanium alloy production (Thesiya et al. 2015). It is mostly used in commercial aircraft jet engines (31%), military aircraft jet engines (20%), Commercial airframes (15%), military airframes (10%), rockets (7%) and helicopters and armaments (1%) and in other industries (16%), with a higher hardness rating of typically around 36–40 on the Rockwell C hardness scale (HRC), and a tensile strength of around 900–1100 MPa, which is comparable to many steels. However, it is important to note that the hardness of titanium alloys can also be influenced by factors such as heat treatment, grain size, and processing techniques, which should be considered during processing (Rocha et al. 2006, Church et al. 2021).

Titanium has, in fact, been considered a potential replacement for steel in certain applications due to its unique properties

(Boyer 1996, 2010) if Ti production costs were to decrease, potentially transitioning from a "future material" to a "common metal" (Froes 2015). Its advantages include (Boyer 1996, 2010) (i) superior strength-to-weight ratio compared to steel, being about 45% lighter than this metal with comparable strength, making it an attractive choice for weight-sensitive applications; (ii) high corrosion resistance; (iii) biocompatibility; (iv) Resistance at high temperatures, retaining strength and integrity under these conditions. Some limitations are, however, also noted, such as (i) costs, as Ti is generally more expensive than steel, primarily due to the higher extraction, processing, and manufacturing costs, as noted previously (ii) Machinability: Titanium is more difficult to machine compared to steel due to its high strength, low elasticity, and low thermal conductivity; and (iii) lower availability due to lower Ti resources. Therefore, for Ti to replace steel, its specific requirements, cost considerations, and the availability of suitable alternatives must all be considered.

Several Ti military applications are also noted. Titanium began being employed in military applications in the 1950s and 1960s by the Soviet Union during the Cold War (Roza 2008), in missiles, submarines and planes, particularly high-performance jets (Jasper 2020). Indeed, the US government considered Ti as such a valuable strategic material during this period that a large stockpile of a porous form of pure Ti was maintained by the Defense National Stockpile Center until the 2000s (Defense National Stockpile Center 2008). Some of the diverse Ti applications in the military include (Fanning 2005) use in aircraft components in alloy form (mainly in structural components, such as airframes, wings, landing gear, and engine parts); in armor and Ballistic protection systems (i.e., body armor plates, vehicle armor, and military vehicle components); in various naval applications (i.e., ship hulls, propellers, valves, heat exchangers, and other critical components that come into contact with seawater, including in submarine pressure hulls and ballast tanks), in the construction of missiles and rockets (rocket casings, nozzles, and other critical components that require strength and heat resistance), in military communication and electronic equipment, in antenna structures, radar systems, satellite components, and military-grade connectors (mostly due to Ti's non-magnetic properties, which reduces electronic signal interference and enhances the reliability of communication systems) and various field equipment for military personnel, such as in lightweight and durable items like knives, canteens, cookware, and other outdoor gear.

Titanium is also used in certain metallurgical processes as a deoxidizer and carbon content reducer (Britannica 2022). When added to molten metals, such as steel or iron, Ti reacts with carbon, forming titanium carbides (TiC), reducing the carbon content in the produced alloy and improving its mechanical properties, such as strength, toughness, and corrosion resistance (Cunat 2004). The effectiveness of Ti as a carbon content reducer depends, however, on several factors, including metal composition, temperature, and specific process conditions. Furthermore, the amount of added Ti and the reaction kinetics plays a crucial role in achieving the desired carbon reduction.

Titanium is also used as a grain refiner, particularly in the production of Al alloys (Britannica 2022), where it is added in small amounts and used to influence the alloy's solidification behaviour, promoting the formation of finer grains, reducing cast defects and facilitating the casting process, improving casting quality (Sigworth 2008). Titanium's effectiveness in this regard, however, depends titanium concentrations, processing parameters, and alloy composition.

Titanium has also been employed in dentistry and medical prosthetic devices since the 1950s, as it is highly biocompatible, due to its resistance to corrosion, also exhibiting high osseointegration capacity, due to its high dielectric constant, insolubility, and the non-permeability of its surface oxide, which does not denature proteins, requiring no adhesives to remain attached, and finally, non-cytotoxicity and bioinertness, making it unreactive to tissue or bone (Raines 2010). These properties make it useful in dental implants, organized implants such as joint replacements, plates, and screws, as well as in the production of dental and orthopaedic instruments such as forceps, drills, screws, and scalpel blades. Additionally, it is employed in dental restorations such as crowns, bridges, and partial dentures, as well as in the creation of maxillofacial prosthetic devices used for facial reconstruction or replacement, including ears, noses, and orbital implants.

The surface of a material plays a critical role in determining how the biological environment responds to artificial medical devices. In the case of Ti implants, Liu et al. (2004) indicate that the standard manufacturing processes often result in an oxidized and contaminated surface layer that is uneven, stressed, and lacks a well-defined structure and not sery suitable for biomedical applications. Therefore, surface treatments become necessary to enhance Ti properties and biocompatibility for medical use. This is conducted through surface biochemical and morphological modifications, leading to decreased bacterial adhesion and inflammatory reactions and increased implant integration and tissue adhesion (Sarraf et al. 2018). For example, various coatings can be applied to Ti surfaces, facilitating better tissue adhesion and integration. These include hydroxyapatite (HA), which promotes bone integration, while bioactive coatings such as calcium phosphate can enhance osteoconductivity (Mohseni et al. 2015). Other methods used to achieve favorable interactions between implant and biological tissues comprise controlled surface roughening through acid etching, sandblasting, or laser, due to increased surface area for bone integration via cell adhesion, proliferation and osseointegration process (Alla et al. 2011). Titanium surfaces can also be modified through functionalization with bioactive molecules, such as growth factors, peptides, or extracellular matrix proteins, enhancing implant surface wettability and bioactivity and improving implant integration with surrounding tissues (Souza et al. 2019). Finally, certain plasma treatments, including cleaning, etching, or polymerization, can be used to modify Ti biomaterial and implant surface chemistry and topography, of titanium implants, biocompatibility and improving cell attachment and tissue integration (Arronson et al. 1997). Figure 2 depicts the most common surface alteration aims for biological applications.

Controlled Ti nanostructures, such as nanodots, nanorods, and nanotubes have also increasingly been applied in regenerative medicine and nanomedicine applications, including in localized

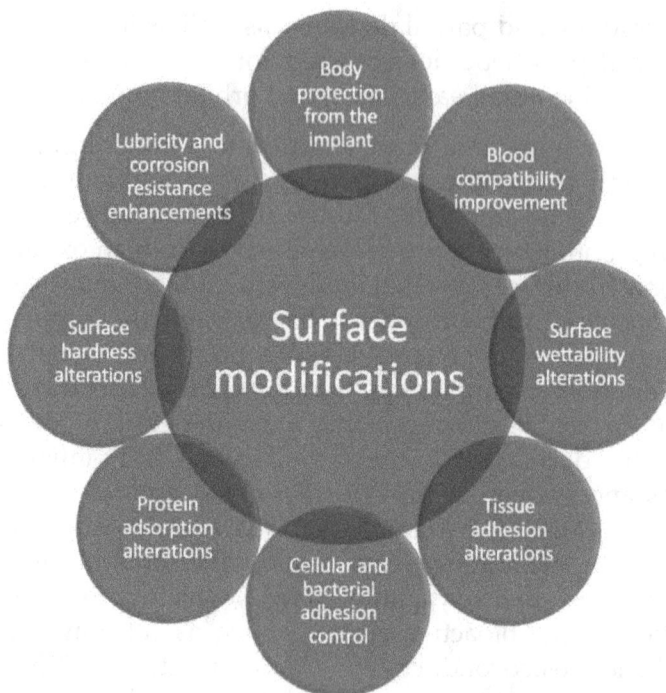

Figure 2. The most common surface alteration aims for biological applications.

drug delivery systems, enabling the loading and release of therapeutic agents, such as drugs or biomolecules in a controlled manner, immunomodulatory agents, antibacterial agents, and hemocompatibility (for a detailed review, see Sarraf et al. 2022).

In one study in this regard, Liu et al. (2023) made use of the immunomodulatory and excellent photothermal properties, as well as selective toxicity to cancer cells, of Ti carbide MXene quantum dots (MQDs) caused by a local surface plasmon resonance effect to investigate their potential applied to the modification of hydroxyapatite (HA) microspheres and hollow MXene quantum dots-modified hydroxyapatite (MQDs-HA) microspheres with controllable shapes and sizes as bone drug carriers. The findings indicated mild Adriamycin storage-release behavior, good pH responsiveness, with accelerated Adriamycin release observed in acidic conditions compared to neutral ones, non-cytotoxic nature assessed by the MTT cytotoxicity assay and, finally, efficient absorption of near-infrared light, converting it into heat, thus conferring pH and near-infrared dual responsiveness to the MQDs-HA carriers. These findings indicate broad application prospects in the field of drug delivery and photothermal therapy.

In another assessment, Ramachandran et al. (2020), successfully synthesized a titanium dioxide/nitrogen-doped graphene quantum dot nanocomposite and assessed it effects on the viability of a breast cancer cell line (MDA-MB-231). The results revealed that Ti doped quantum dot nanocomposite exhibited no toxicity towards breast cancer cells at concentrations up to 0.1 mg mL^{-1}. Furthermore, even at higher concentrations (0.5 and 1 mg mL^{-1}), the nanocomposite displayed significantly lower cytotoxicity when compared to pristine TiO$_2$. The authors conclude that the synthesized compound holds promise for further investigation as a novel TiO$_2$-based nanomaterial in biomedical applications, particularly as an anti-cancer strategy.

Another study synthesized oxide nanotubes (ONTs) on a Ti–6Al–7Nb Alloy and evaluated its potential as a potential drug carrier (Łosiewicz et al. 2021). The authors applied anodic oxidation to modify the alloy surface of a Ti–6Al–7Nb bar and evaluated gentamicin sulfate loading and release kinetics. Gentamicin sulfate is a broad-spectrum aminoglycoside antibiotic routinely employed against bacterial infections, particularly valuable in the treatment of severe bone tissue infections induced by several bacteria. The authors indicate that the utilization of gentamicin sulfate in controlled drug release systems offers the advantage of precise drug delivery directly to the tissues surrounding the implant and that this targeted approach enhances the immune system's response and hinders the proliferation of bacteria. The developed ONTs obtained exhibited high capacity as drug-eluting compounds, with a controlled and extended gentamicin sulfate release from the ONTs within a phosphate-buffered solution, offering a means to circumvent oral supplementation while effectively mitigating inflammation following implantation.

The Case of Ti Nanoparticles

Nanoparticles (NP) are defined as particles of matter with at least one dimension in the nanoscale range, ranging between 1 and 100 nm in diameter, although this definition is sometimes extended to larger particles up to 500 nm (USEPA 2023). Nanoparticles often exhibit different properties from their bulk counterparts, due to their smaller size and significantly higher surface area-to-volume ratio, leading to increased reactivity (Guisbiers et al. 2012). The most common NP used across various industries include silver, gold, iron, zinc, and aluminum and Ti NP (Schrand et al. 2010).

Most extracted titanium (about 90%) is converted to TiO_2, one of the most commercially important Ti compounds mainly sourced from brookite, octahedrite, anatase, and rutile (Britannica 2022). This form of Ti comprises a white powder extensively employed as a pigment in paints, enamels, and lacquers, as well as in the food industry (Britannica 2022).

The annual worldwide production of titania powder was estimated at around 5 million tons in 2009, with about 2.5% of this total comprising nano-sized transparent titania compounds produced in the rutile and anatase forms, termed Ti nanoparticles (TiO_2 NP) (Robichaud et al. 2009). In 2022, the global titania dioxide powder market was estimated at about 21 billion USD, projected to reach 32 billion USD by 2030 (Vantage Market Research 2023).

In 2015, TiO_2 NP production increased to about 10% of the total titania powder produced worldwide (Robichaud et al. 2009). These Ti nanoparticles are one of the three most produced nanomaterials, alongside silicon dioxide and zinc oxide NPs (Keller et al. 2013, Zhang et al. 2015), thus considered commodity chemicals.

Commodity chemicals, also known as bulk chemicals, refer to chemicals that are produced and traded in large quantities in large-scale manufacturing processes to meet widespread industrial or consumer demands and to satisfy global markets and used as raw materials or intermediates in various industries (Kannegiesser 2008).

The production of Ti NPs typically involves specialized techniques and processes to achieve the desired size and characteristics, with careful control over various parameters to ensure desired size, shape, and properties (Ahmed et al. 2023). The main ones are (Nyamukamba et al. 2018): (i) Chemical Precipitation, where Ti salts are dissolved in a solvent, and a reducing agent is added to initiate the precipitation of Ti nanoparticles. The size and properties of the nanoparticles can be controlled by adjusting the reaction conditions, such as temperature, pH, and the choice of reducing agent; (ii) the Sol-Gel Method, which involves the synthesis of Ti nanoparticles through the hydrolysis and condensation of titanium alkoxide precursors. The process includes the formation of a sol (colloidal suspension) followed by gelation and subsequent drying or calcination to obtain Ti nanoparticles; (iii) Hydrothermal Synthesis, which involves the reaction of Ti precursor materials mixed with a solvent under high-pressure and high-temperature conditions in a sealed container to promote the formation of Ti nanoparticles. This method allows for control over particle size, morphology, and crystallinity; (iv) plasma sputtering, where a high-energy plasma is used to bombard a titanium target under controlled plasma parameters and deposition conditions, causing the ejection and deposition of Ti atoms onto a substrate and (v) mechanical milling, a solid-state method where Ti powders are subjected to intense mechanical forces (milling or grinding), to break down the bulk material into NP, producing NP with controlled size and structure.

Titanium dioxide NPs have a relatively positive public acceptance, as they are widely considered a "natural" material, which may be misleading (Skocaj et al. 2011). Their long use has, in fact, led TiO_2 NP to be categorized as a legacy nanomaterial (USEPA 2010). The term "legacy nanomaterial" is used to refer to nanomaterials that have been widely produced and used before the establishment of specific regulations or guidelines for nanomaterial safety (USEPA 2010). These materials were and still are introduced into various products and applications without thorough understanding of their potential health and environmental impacts. The classification of a nanomaterial as a legacy nanomaterial does not imply that it is inherently hazardous, but rather reflects the need for further assessments and understanding its potential risks aiming at appropriate risk management strategies. Unfortunately, however, as stated by Hristozov et al. (2012), most approaches to date only serve as preliminary risk screening and/or research prioritisation tools, not intended to support regulatory decision making, indicating the clear need for further assessments in this regard.

Titanium dioxide NP are also categorized as a nanomaterial of emerging concern (USEPA 2010), even though the term "contaminants of emerging concern" is usually associated with new synthetic molecules recently identified as environmental contaminants (Khan et al. 2022). In this case, the term 'emerging' follows the definition given by Batley and Campbell (2002), referring not to Ti itself, but instead to the new and emerging uses of this element and the deleterious effects and risks it may pose to human health and the environment. As noted above for legacy nanomaterials, it is important to note that the classification of a nanomaterial as an emerging concern does not necessarily imply that it is inherently hazardous but, instead, highlights the need for further research, risk assessment, and the implementation of appropriate safety measures to ensure the responsible development and use of nanotechnology (Poynton and Robinson 2018).

Nanoparticle characterizations enable the precise tuning of TiO_2 NPs for specific applications. For example, understanding the shape and morphology of TiO_2 NPs is vital, as these parameters may significantly affect their reactivity and performance. Surface charge assessments, measured through zeta potential analysis, are crucial for assessing colloidal stability of TiO_2 NPs in suspensions, as this parameter impacts dispersion, aggregation, and interactions with other materials. Furthermore, stability assessments ensure that the nanoparticles retain their properties and performance over time, preventing agglomeration or degradation. In addition, determining particle size and size distribution of TiO_2 NPs is essential in applications where uniformity is critical, as different particle sizes may display varying optical, electronic, and catalytic properties. Understanding the surface properties, including the presence of functional groups (e.g., hydroxyl, carboxyl) or surface modifications, helps tailor TiO_2 NPs for specific applications, such as drug delivery, where surface functionalization can improve biocompatibility and targeting. Concerning doping, adding other elements, compounds or drugs to TiO_2 NPs surfaces can modify their electronic structure and enhance their performance in various applications, such as drug release. Finally, purity assessments ensure that the investigated materials do not contain impurities or contaminants that could adversely affect their performance or safety. Figure 3 depicts the most common TiO_2 NP physicochemical characterizations for various applications.

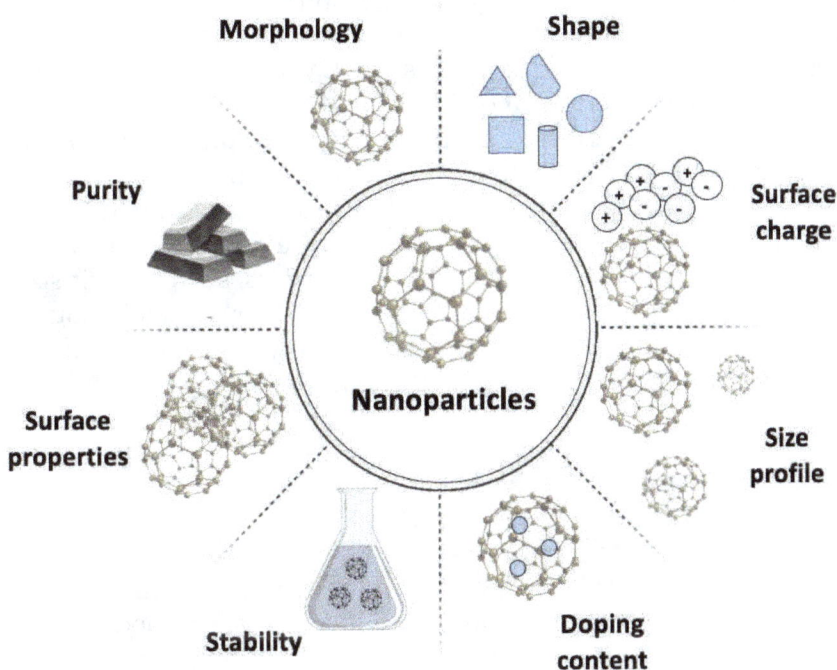

Figure 3. The most common TiO_2 NP physicochemical characterizations for their various applications.

Titanium Dioxide Nanoparticle Applications

Titanium oxide NP, are now highly applied to common household and daily life products worldwide, mainly due to (i) high photocatalytic properties, making them useful in Advanced Oxidative Processes (AOPs) (Wold 1993); (ii) increased UV absorption, effectively absorbing and scatter UV radiation, which makes them useful as UV-blocking agents in sunscreens, cosmetics, coatings, and materials for UV protection, as well as food additive (Egerton and Tooly 2004); (iii) antibacterial and antimicrobial properties, suitable for applications in healthcare settings, antibacterial coatings, and water treatment (Chen et al. 2019) and (iv) biomedical applications, such as drug delivery, bioimaging, and tissue engineering (Jafari et al. 2023). However, the specific characteristics of TiO_2 nanoparticles, such as their size, surface charge, and surface reactivity, can influence their biological interactions, and must, therefore, be carefully controlled. Figure 4 displays the main current TiO_2 NP applications.

- *Titanium oxide nanoparticle in paints and inks*

In its TiO_2 form, Ti is extensively applied in the manufacture of white pigments as "titanium white" or Pigment White 6, applied to coatings, plastics and paint to scatter light, as it absorbs almost no incident light in the visible spectrum (Pitre et al. 2017). Furthermore, TiO_2 exhibits excellent resistance to chemical attacks, good thermal stability and resistance to ultraviolet (UV) degradation, also displaying a very high refractive index of 2.70 for rutile TiO_2 and 2.55 for anatase TiO_2. As a result, relatively low pigment levels are required to obtain a white opaque coating (Gázquez et al. 2014).

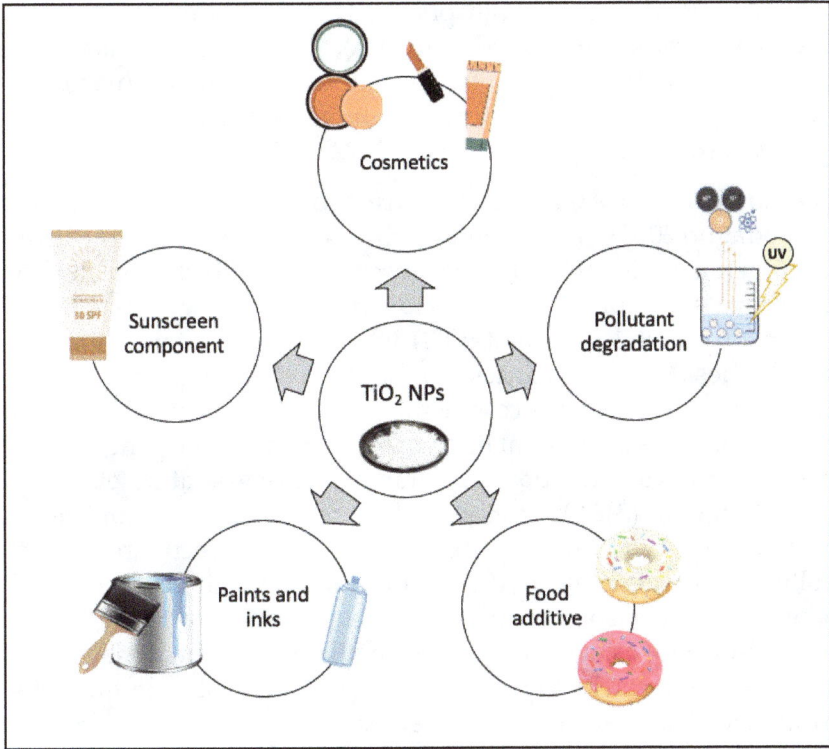

Figure 4. Main titanium oxide nanoparticle applications.

This compound is also applied to printing inks, paper and textiles due to its brightness, and in the pharmaceutical and cosmetic industries to obtain appropriate colours for marketing objectives (Gázquez et al. 2014).

- *Titanium oxide nanoparticle use in Advanced Oxidative Processes (AOPs)*

Titanium oxide also comprises a popular photocatalyst used in Advanced Oxidative Processes (AOPs). Advanced Oxidative Processes consist in treatment technologies aimed at degrading and mineralizing recalcitrant organic matter and environmental contaminants present in wastewater through hydroxyl radical ($^\bullet$OH) reactions, including pesticides, dyes, pharmaceuticals, and other metals, due to its high photoactivity (Saggioro et al. 2014). These processes are termed "advanced" because they are more powerful and effective, as well as more diverse, than conventional treatment methods, resulting in more efficient contaminant removals (Bracamontes-Ruela et al. 2022).

The most common AOPs currently in use today include the following (Capodaglio 2020): (1) the Fenton process, which involves the use of hydrogen peroxide (H_2O_2) in the presence of a catalyst, typically iron or iron salts (Fe^{2+} or Fe^{3+}) under acidic conditions, to generate highly reactive hydroxyl radicals ($\bullet OH$) which break contaminants into simpler, less toxic substances; (2) the Photo-Fenton process, which combines the Fenton process with ultraviolet (UV) or visible light irradiation. This additional light energy enhances the generation of hydroxyl radicals, thereby accelerating the degradation efficiency of pollutants; (3) Ozonation, which involves the application of ozone (O_3), a powerful oxidant that can directly react with pollutants or generate hydroxyl radicals through reactions with other oxidizable species to oxidize and degrade pollutants and; (4) Photocatalysis, which utilizes semiconductor materials, such TiO_2, as catalysts in the presence of UV or visible light. When irradiated, the catalyst generates electron-hole pairs, leading to the production of reactive species like $\bullet OH$, which can degrade organic pollutants. A summary of the main AOPs is depicted in Figure 5.

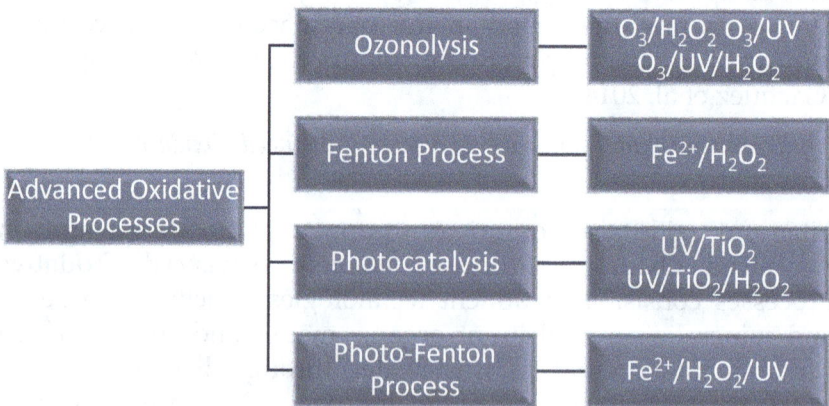

Figure 5. Summary of the main Advanced Oxidative Processes and their catalysts.

Titanium dioxide is commonly used in Advanced Oxidation Processes (AOPs) due to its photocatalytic properties and the fact that it is relatively inexpensive (Stasinakis 2008). It acts as a catalyst by absorbing photons from light sources, which excites electrons and creates electron-hole pairs, which then generate ROS, which have strong oxidizing capabilities. The presence of TiO_2 NP, either dispersed in the water or immobilized onto surfaces, such as porous media or catalyst supports, enhances AOP efficiency by providing a larger surface area for photocatalytic reactions to occur (Navidpour et al. 2023). Furthermore, TiO_2 NP photocatalytic activity can be easily controlled by adjusting factors such as particle size, morphology, and doping. However, challenges exist in optimizing the efficiency of TiO_2-based AOPs, including addressing issues such as catalyst recovery, agglomeration, and the need for UV light sources (Chen et al. 2020a).

In this regard, one study was able to degrade several pharmaceuticals (alprazolam, clonazepam, diazepam, lorazepam, and carbamazepine) at significant rates of 60 to 80% employing TiO_2-photocatalysis in the form of TiO_2/UV-A and TiO_2/H_2O_2/UV-A AOPs (Bosio et al. 2019). This was performed under both sunlight and artificial irradiation conditions in ultrapure water, decreasing to about 50% for both processes when employing a WWTP effluent. The combination of both processes was shown to be an adequate approach, increasing drug removal up to 90%. Another assessment employed supported TiO_2 solar photocatalysis at semi-pilot scale applied to pesticide degradation of pesticides found in citrus processing industry wastewater (Jiménez-Tototzintle et al. 2015), reporting good mechanical stability and photocatalytic activity for the TiO_2 supported catalyst, even in complex matrices for several pesticides in a real municipal wastewater treatment plant secondary effluent. In yet another study, TiO_2-photocatalysis was applied to the remediation of model textile wastewaters containing azo dyes, an important source of environmental contamination (Saggioro et al. 2011). Following adequate dye degradation, unlike most studies, the authors commented on the possibility of TiO_2 catalyst recycling for use in other reactions. However, although AOPs are considered clean technologies, capable of achieving high organic compound degradation rates (Bosio et al. 2019), the ubiquitous use of TiO_2 ostensibly for contaminant treatment, however, also presents significant risks. This is due to the fact that, TiO_2 recovery from these processes is extremely difficult (Gopinath et al. 2020), mainly because TiO_2 NP used in AOPs are often in the nanometer range, and their small size and high surface area make it challenging to separate and recover them efficiently, resulting in secondary water body contamination, requiring additional separation and purification steps, which can add complexity and cost to the treatment process (Cheng et al. 2014, Teng et al. 2014). Furthermore, TiO_2 NP can aggregate or agglomerate, forming larger particles or clusters, which can further complicate their recovery (Loosli et al. 2013). Thus, stable dispersion and preventing aggregation are key factors in successful recovery. In addition, the presence of other chemicals, suspended solids, or organic compounds in the water matrix can interfere with TiO_2 nanoparticle recovery, as they may interact with TiO_2 or hinder separation processes (Liriano-Jorge et al. 2014).

To overcome these challenges, various separation and recovery techniques have been explored, such as filtration, sedimentation, centrifugation, and advanced separation technologies like membrane filtration or magnetic separation (Liriano-Jorge et al. 2014). However, no universally standardized method for TiO_2 NP recovery from AOPs has been established to date, and the choice of recovery method depends on the specific application, water quality, and economic feasibility.

- *Titanium oxide nanoparticle use in cosmetics*

Titanium oxide has been applied in many different cosmetics, including but not limited to, day creams, foundations, and lip balms, since the 1980s, aiming at providing protection against the carcinogenic effects of UV radiation (Dréno et al. 2019), due to Tis high refractive index. Up until recently, several studies have been conducted to date regarding the toxicity of cosmetic-incorporated Ti. Concerning skin exposure, assessments in both humans and animals indicate that Ti NPs cannot penetrate outer skin levels of viable cells, thus not resulting in systemic exposure (Dréno et al. 2019).

However, the Scientific Committee on Consumer Safety indicates the need for caution with regard to lung exposure, as lung inflammation has been reported in animal studies, thus, not recommending the use of TiO$_2$ NPs in sprayable products (Dréno et al. 2019).

Some studies have been carried out in this regard, although they are still scarce. For example, one study aiming to assess potential TiO$_2$ NP toxicity in cosmetics assessed a 1.0 g L^{-1} suspension in a simulated sweat solution applied on Franz cells for 24 h using intact and needle-abraded human skin (Crosera et al. 2015). The cytotoxicity of TiO$_2$ NP was assessed using MTT, AlamarBlue®, and propidium iodide (PI) uptake assays (an index of necrotic cell death) on HaCat keratinocytes for different exposure durations (24 h, 48 h, and seven days). After 24 h of exposure, no detectable Ti levels were observed in the test solutions, applied to either intact or damaged skin. This element was, however, detected in the epidermal layer (0.47 ± 0.33 µg cm^{-2}) after 24 h of exposure, while the concentration in the dermal layer remained below the mehtod limit of detection. In the case of damaged skin, the overall concentration was similar (0.53 ± 0.26 µg cm^{-2}). Cytotoxicity studies on HaCaT cells revealed that TiO$_2$ NP only induced cytotoxic effects at very high concentrations, reducing cell viability after seven days of exposure. The authors indicate that these findings demonstrate that TiO$_2$ NP cannot penetrate intact or damaged skin, and they are only present in the stratum corneum and epidermis. Additionally, the low cytotoxic effects observed on human HaCaT keratinocytes suggest that these nano-compounds may only exhibit potential toxic effects on the skin after prolonged exposure.

In a study conducted on Yucatan micropigs (Senzui et al. 2010), the skin penetration of four different types of TiO$_2$ NP was examined, namely T-35 (non-coating, 35 nm), TC-35 (with almina/silica/silicon coating, 35 nm), T-disp (mixture of almina-coated and silicon-coated particles, dispersed in cyclopentasiloxane, a volatile silicone fluid commonly used in cosmetics, 10 × 100 nm), and T-250 (non-coating, 250 nm). Various skin conditions were considered, such as intact skin, stripped skin, and hair-removed skin, to investigate the impact of dispersion and skin condition on penetration.

The suspensions were applied to the skin for 24 h, followed by cyanoacrylate stripping. Titanium concentrations in skin were determined by ICP-MS, with the, T-35 and T-250 exhibiting a tendency to aggregate in the suspension, whereas TC-35 and T-disp demonstrated good dispersion properties. Regardless of the type of TiO_2 NP, no penetration was observed in both intact and stripped skin. However, when TC-35 was applied to hair-removed skin, a significantly higher concentration of Ti was detected, and scanning electron microscopy-energy-dispersive X-ray spectroscopy analyses revealed Ti penetration into empty hair follicles located more than 1 mm below the skin surface, although no penetration in the dermis or viable epidermis.

Studies concerning Ti contents in cosmetics have also been conducted. In one assessment, several metals, including Ti, were determined in twenty-two lipstick products available in Saudi Arabia (Zainy 2017). Titanium in these products is permitted by the FDA, used as a whitening agent, to soften reds into pinks, and as an antioxidant (Zainy 2017). In that study, Ti contents averaged 46.59 ± 0.109 $\mu g\ g^{-1}$. These data are important as baseline data for this type of assessment, even though no safe limits have been established for Ti in cosmetic products in Saudi Arabia, or even in many other countries. The authors, however, indicate that prolonged use of products containing Ti may pose health risks to humans, especially as women have been estimated to ingest about 1.8 kg of lipstick in their lifetime without intention (Patel 2016).

In another assessment, Lee et al. (2020) examined the association between urinary oxidative stress markers and exposure to cosmetics containing TiO_2 and ZnO NPs in 40 cosmetics salesclerks and 24 clothing salesclerks, categorized based on their exposure to ZnO and TiO_2 NPs.

Concerning Ti, a total of 20 cosmetics samples were analyzed, with 19 meeting the European Union's definition of TiO_2 NPs, while 15 samples met the definition for ZnO nanomaterials. Participants with a higher co-exposure index to ZnO and TiO_2 NPs exhibited significantly higher baseline levels of urinary 8-hydroxy-2'-deoxyguanosine (8-OHdG) concentrations compared to the lower co-exposure group. After adjusting for potential confounding factors, the authors observed a significant positive association between the TiO_2 and ZnO NP co-exposure index and both urinary 8-OHdG baseline concentrations and creatinine-adjusted concentrations, postulating that there is a likelihood of increased urinary 8-OHdG levels associated with the use of sunscreens containing NP.

- *Titanium oxide nanoparticle use in the food industry*

Titanium oxide is also extensively applied in the food industry as a food additive, in the form of food grade titanium dioxide (TiO_2), commercially known as E171 in Europe and as INS171 in the USA (Ropers et al. 2017). The initial approval for the use of E171 took place in 1966 by the United States Food and Drug Administration (FDA), followed by the European Union in 1969, both based on the Codex Alimentarius guidelines established by the Food and Agriculture Organization/World Health Organization (FAO/WHO) (Ropers et al. 2017).

Food additives are intentionally added to food during the manufacturing, processing, preparation, treatment, packaging, transport, or storage steps of foodstuffs (Branen et al. 2001). Color additives, as in the case of E171, are used to improve food appearance, enhancing its visual appeal, aiming at better customer acceptance (Boutillier et al. 2022). In the case of E171, this compound is used to make food more visually appealing, due to its brightness, high refractive index, and resistance to discoloration (Weir et al. 2012).

The average size of food-grade TiO_2 used to be composed of relatively coarse particles. This, however, is no longer the case, as smaller mean particle sizes and increasing amounts of E171 nanoparticles in food-grade TiO_2 are now noted due to improved production technologies (Chen et al. 2020b). This has led to increased human dietary and environmental exposure to TiO_2 NP and ensuing human health concerns.

The main food categories contributing to E171 dietary exposure are stated as "fine bakery wares, soups, broths, and sauces (for infants, toddlers, and adolescents); and soups, broths, sauces, salads and sandwich spreads (for children, adults and the elderly). Processed nuts are also a main contributing food category for adults and the elderly" by the EFSA (EFSA 2023). Some authors indicate candies, sweets and chewing gums as foods containing the highest TiO_2 contents, which may lead to higher risks for children, which usually consume these items in higher amounts than adults and exhibit lower body weights and developing immune systems (Weir et al. 2012). The use of TiO_2 was authorized as a food additive in the EU according to Annex II of Regulation (EC) No 1333/2008, where it should contain at most 50% of particles in the nano range to which consumers may be exposed.

Acceptable daily intake levels (ADI) for TiO_2 as a food additive were established by several countries, as not enough evidence of toxicity was available for these compounds in humans up until a few years ago. Recently, however, the European Food Safety Authority (EFSA) has recently updated its safety assessment of TiO_2 as a food additive following a request by the European Commission in March 2020, as the previous assessment published in 2016 highlighted the need for more research to fill data gaps. This additive is, therefore, now no longer considered safe according to EU Commission Regulation (EU) 2022/63 (EFSA 2023), as genotoxicity due to ingestion could not be excluded since, despite a low absorption of TiO_2, these compounds can accumulate in the body. Consequently, a safe level for daily intake could not be established. According to the new regulation, food and drinks containing TiO_2 can be placed on the market until 7 August 2022 and can remain on the market until their expiry date. After that date, TiO_2 as a food additive will no longer be allowed in the manufacture of food items in the European Union or for imports into the European market. Other countries are expected to follow suit and also ban E171.

The routine use of TiO_2 NPs in our society leads to exorbitant volumes of these nanoparticles to be continuously discharged through industrial and domestic effluents into aquatic environments around the globe, as most wastewater plant treatments are not equipped to deal with NP (Sirés and Brillas 2012). In fact, TiO_2 has been postulated as reaching highest concentrations in surface waters among other commonly employed nanomaterials, posing a significant threat to aquatic ecosystems (Gottschalk et al. 2009, 2010). A recent study, for example, estimated that between 4.4 and 32.4 mg of TiO_2 from personal care products are released into wastewater systems per household in the USA each day, with about 0.15 to 2.06 µg L^{-1} of this daily input persisting in WWTP effluents after treatment (Wu et al. 2020). This, of course, has led to significant environmental and Public Health concerns, as TiO_2 NP has been shown to undergo partitioning in both the abiotic and biotic compartments, accumulating, for example, in the surface microlayer of water bodies at concentrations ranging from 1 to 1000 µg L^{-1} and aggregating and thus, sinking and settling into the sediment (Botta et al. 2011, Labille et al. 2020), as well as being incorporated into the biota. These processes can influence their behavior, distribution, and potential effects on aquatic organisms.

Analytical techniques to determine TiO_2 NPs, however, are still incipient, as traditional metal analyses methods (i.e., acid digestion and subsequent analysis by flame atomic absorption spectrophotometry (F-AAS), or inductively coupled plasma optical emission spectroscopy (ICP-OES) or inductively coupled plasma mass spectroscopy (ICP-MS)), are usually ineffective for TiO_2 NPs, resulting in poor recovery and reproducibility (Shaw et al. 2013). Some of the reasons for this are discussed in Shaw et al. (2013) and include the fact that traditional spectrophotometry assumes the metal is dissolved in the analysed sample, and that TiO_2 NPs are resistant to nitric acid digestion, leading to concerns on NP losses by aggregation within the instruments and inadequate sample nebulization, both contributing to poor detection. These authors also indicate that aggressive digestion methods employing, for example, hydrofluoric acid and hot concentrated sulfuric acid, TiO_2 NPs are hazardous and time consuming, and make further analyses, such as particle number quantification and even simple presence/ absence particle estimates impossible. Thus, the authors state that new methods are required, such as single particle inductively coupled plasma mass spectrometry (SP-ICP-MS) for the analysis of NPs in water samples, or both ICP-OES and ICP-MS following adequate sample treatment for this specific end. Other analytical techniques for Ti determination include the following, according to Gaafar et al. (2022): (i) Transmission Electron Microscopy (TEM), which provides high-resolution images of the particles, allowing for size, shape, aggregation and morphology analyses, (ii) Scanning Electron Microscopy (SEM), providing detailed surface morphology information, (iii) energy-dispersive X-ray spectroscopy (EDS), which can be coupled with SEM to analyze elemental particle composition, (iv) Dynamic Light Scattering (DLS), used to measure the hydrodynamic size distribution, polydispersity, and stability of TiO_2 NPs in liquid suspension. This method can also be combined with zeta potential measurements to assess the surface charge of the particles, (v) X-ray Diffraction (XRD), which analyses crystal structures and is able to identify different TiO_2 crystal phases, (vi) Fourier Transform Infrared Spectroscopy (FTIR), able to identify TiO_2 NPs through their characteristic infrared absorption bands, also capable of detecting NP surface modifications or coatings on the nanoparticles and, finally, the simpler technique of Ultraviolet-Visible Spectroscopy (UV-Vis), commonly used to analyze optical NP properties of nanoparticles by assessing characteristic absorption peaks in the UV region due to their bandgap properties.

Titanium in Aquatic Systems

Titanium in aquatic ecosystems can originate from different sources, including weathering of Ti-containing minerals, river and stream runoff and industrial and anthropogenic activities. This element can exist in different forms in aquatic environments, including dissolved ions (Ti^{2+} and Ti^{4+}), suspended particulate matter, and sediment-bound forms (Linnik and Zhezherya 2015). The behavior of Ti in water, as well as many other metals, is, however influenced by various factors such as pH, redox conditions, complexation with organic and inorganic ligands, temperature, oxygen content, ionic strength, and the structure and concentration of the natural organic matter (NOM), as well as interactions with suspended particles or sediment, leading to differential bioaccumulation patterns (Freixa et al. 2018).

Concerning TiO_2 NPs, Keller et al. (2010) conducted a study demonstrating the variability of aggregation and deposition of TiO_2 NPs in different natural and artificial water systems, reporting that TiO_2 NPs aggregation and deposition rates were significantly influenced by ionic strength and total organic carbon levels, particularly in seawater samples. TiO_2 NPs concentrations also played a role, with aggregate sizes reaching approximately 1 μm at 10 mg TiO_2 NPs L^{-1} and up to 2 μm at 50 and 200 mg TiO_2 NPs L^{-1}. Water temperature and flow velocity were identified as key factors affecting the aggregation process and sedimentation behavior of n-TiO_2, as highlighted by Lv et al. (2016). Additionally, it has been observed that salinity facilitates nanoparticle aggregation (Jiang et al. 2009). Doyle et al. (2015) further investigated the sedimentation behavior of three forms of TiO_2 NPs in natural seawater. Their findings revealed that over 92% of the initial mass of the nanoparticles had settled after 72 h of mixing, followed by a 45-minute settling period. Heteroaggregation of TiO_2 NPs in the presence of clay has also been documented (Labille et al. 2015, Adam et al. 2016). Moreover, the stability of these heteroaggregates was found to be significantly influenced by the presence of humic acid (Wang et al. 2015).

Another assessment evaluated the disagglomeration of TiO_2 NPs aggregates in the presence of alginate and Suwannee River humic acids at varying concentrations, representing real environmental conditions (Loosli et al. 2013). To assess stability, dynamic light scattering and electrophoretic measurements were conducted, focusing on typical environmental concentrations of natural organic matter and a pH value corresponding to the point of zero charge of TiO_2 NP. Under those conditions, the surface charge of TiO_2 NP was neutralized, allowing for the formation of large agglomerates. Alginate and Suwannee River humic acids, exhibiting negative structural charges at this pH level, adsorb onto the surface of the nanoparticle agglomerates, leading to the deagglomeration and significant redispersion of TiO_2 NP into fragments. The authors indicate that their findings highlight the importance of electrostatic forces and steric interactions in the disagglomeration process, and that the physicochemical properties of natural organic matter were found to influence the kinetics and significance of TiO_2 NP fragmentation during disagglomeration, which may have implications for TiO_2 NP remediation in natural waters. Furthermore, they indicate that data demonstrate that the presence of natural organic matter at typical environmental concentrations induces substantial disagglomeration of large submicron nanoparticle agglomerates, of significant relevance when assessing the risks associated with manufactured nanoparticles, as it highlights the potential transformations of micron-sized structures that these nanoparticles can form.

Biological Titanium Effects

Although some of the first Ti toxicity assessments classified this metal as biologically inert (Ophus et al. 1979, Lindenschmidt et al. 1990), increased human and environmental exposure to Ti compounds, especially TiO_2 NP, have led to toxicological and ecotoxicological investigations in this regard (Skocaj et al. 2011). This led to further toxicological and ecotoxicological assessments, which resulted in TiO_2 reclassification as a class 2B compound, possibly carcinogenic to humans, by means of inhalation by the International Agency of Research on Cancer, a World Health Organization (WHO) agency that evaluates the carcinogenicity of various substances based on available scientific evidence, due to sufficient evidence of cancer risk in animals (IARC 2006). Furthermore, TiO_2 in its powder form has also been recently classified as a category 2 suspected carcinogen by inhalation by the European Union (EU 2020b). The classification of a compound as a Class 2B carcinogen does not, however, mean that exposure will automatically result in cancer, as this classification is based on potential risks and should be considered in a context of exposure levels, duration, and other factors (Boobis et al. 2016).

Although Ti tends to be present in low concentrations in water and sediment, and its uptake by organisms is generally low, several *in vitro* and *in vivo* studies have indicated deleterious toxic Ti effects, mainly in the form of TiO_2 NP, which include, but are not restricted to, oxidative stress and cellular damage processes (Reeves et al. 2008, Hao et al. 2009, Chen et al. 2014, Tay et al. 2014), leading to cytotoxic, physiological and genotoxic effects following both short- and long-term exposure (Shi et al. 2013, Tassinari et al. 2014).

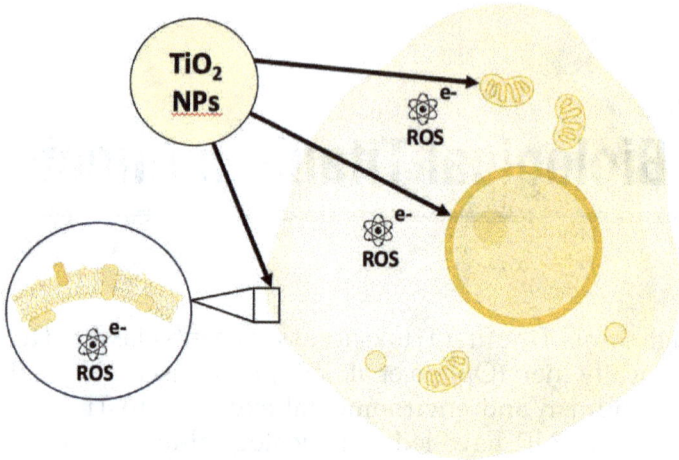

Figure 6. Representative TiO$_2$ nanoparticle interactions with cell organelles and the double phospholipid membrane, resulting in the generation of Reactive Oxygen Species (ROS).

Furthermore, even though Ti toxicity assessments are still in their infancy, the main Ti mechanisms of action so far have been noted as threefold, comprising the generation of Reactive Oxygen Species (ROS), leading to intracellular oxidative stress, electrostatic attachment to cell membranes, resulting in cell wall damage and lipid peroxidation and bonding to intracellular organelles and biological macromolecules, resulting in compromised membrane and leading to further detrimental effects (Hou et al. 2019) (Figure 6).

The potential for TiO$_2$ NPs to induce oxidative stress, however, depends, of course, on various factors, including their physicochemical properties (size, shape, surface characteristics), dose, exposure duration, concentration, and the specific biological system involved. Furthermore, the extent of oxidative stress and resulting effects can vary across different studies and models. Ecotoxicological assessments are being increasingly carried out to unravel these and other questions concerning Ti risks to the environment and living organisms, through laboratory experiments, field studies, and modeling approaches to assess the risks of contaminants to ecosystems, populations, and individual organisms. The findings help guide environmental management practices, inform policy decisions, and contribute to the protection and conservation of ecosystems and biodiversity.

Some of these studies will be discussed below, categorized according to taxonomic group. Although most studies focus on aquatic organisms, some terrestrial taxa have also been the subject of ecotoxicological assessments, and studies employing both *in vitro* and *in vivo* models, as well as in humans are also available. Both *in vitro* and *in vivo* models are important in examining Ti, each exhibiting their own particularities and able to further knowledge on different effects, such as organ function, cellular mechanisms, and systemic responses. This in turn, will aid researchers in understanding Ti toxicokinetics, i.e., absorption, distribution, metabolism, and excretion and, consequently, assessing potential ecological and human health risks (Magdolenova et al. 2014).

Phyto- and zooplankton

Research on the effects of TiO_2 NP on both phyto- and zooplankton can provide valuable insights into their potential ecological consequences, as both these groups comprise the base of many trophic food chains in aquatic ecosystems worldwide. This will, in turn, help identify potential indirect effects on higher trophic levels, which may result in series of indirect interactions, with changes in one species or trophic level leading to subsequent changes in other species or trophic levels, jointly termed cascade effects (Ripple et al. 2016).

Phytoplankton comprise the base of oceanic food webs, consisting in the dominant primary producers in marine ecosystems (Behrenfeld et al. 2006) They are also an integral part of both the global carbon cycle, as well as other biogeochemical cycles. Thus, disturbances to these components may result in potential coastal marine food web imbalances, leading to significant negative effects on the ecosystems that they support (Miller et al. 2012). Disturbances in this regard include, but are not limited to nutrient limitation (i.e., eutrophication, leading to harmful algal blooms and oxygen-depleted zones), temperature changes, light availability, ocean acidification and water circulation changes, leading to impacts on higher trophic levels, carbon cycling, and overall ecosystem productivity (Sigman and Hain 2012, Litchman et al. 2015, Hoppe et al. 2018, Fernández-González et al. 2022).

Exposure to TiO_2 NP has been shown to result in toxicity in marine phytoplankton phytoplankton representing three major groups, diatoms (Phylum: Heterokontophyta, Class: Bacillariophyceae), green chlorophytes (Phylum: Chlorophyta, Class: Chlorophyceae), and prymnesiophytes (Phylum: Haptophyta, Class: Prymnesiophyceae) in seawater under natural and relatively low ultraviolet light levels near the ocean's surface (< 1 m depth in coastal waters) (Miller et al. 2012). Toxic effects were detected at the lowest tested concentration of 1 mg L^{-1} in *Isochrysis galbana*, indicating a no-effect concentration at less than 1 mg L^{-1}. In the other two tested species, *Thalassiosira pseudonana*, and *Dunaliella tertiolecta*, significant toxicity was evident at 3 mg L^{-1}, although a slight growth rate decrease was noted for *D. tertiolecta* at 1 mg L^{-1}. Increased radical oxygen species (ROS) production was detected with increasing TiO_2 NP concentrations, and no effects were noted when UV light was blocked. A scanning electron microscopy analysis indicated that the TiO_2 NP adhered to phytoplankton cell surfaces. The authors indicate the findings as of significant concern, as phytoplankton are the most important primary producers on Earth and toxic TiO_2 NP effects may cause decreased marine ecosystem resiliency.

In another study, a 2-week microcosm experiment with a natural freshwater bacterial community assessed TiO_2 NP effects at 0, 1, 10 and 100 mg L^{-1} on planktonic and sessile bacteria under dark conditions (Jomini et al. 2015). The findings indicate an increase of planktonic bacterial abundance at the highest TiO_2 NP concentration of 100 mg L^{-1} concomitant with a decrease in sessile bacteria, while overall diversity was found to be lower for planktonic bacteria and higher for sessile bacteria at this concentration and decreased Comamonadaceae was observed alongside an increase of Oxalobacteraceae and Cytophagaceae among the planktonic communities. The authors indicate that these findings indicate the need to investigate potential synergic/interactive effects between TiO_2 NP and planktonic bacteria and biofilms in higher trophic levels in a risk assessment content, due to potential impacts on freshwater food webs and ecosystem functioning.

Another assessment indicated severe TiO_2 photocatalytic effects on the marine plankton *Chattonella antiqua*, a red tide flagellate (Matsuo et al. 2001). Following UV irradiation for 1 hour, body deformations from spindle to rounded were observed within 20 minutes, and, while the deformed *C. antiqua* recovered to a normal shape when maintained in the same conditions without UV irradiation for more than 40 minutes, prolonged UV irradiation of over 100 min led to burst cells and death.

Zooplankton also play a crucial role in aquatic ecosystems, comprising a vital link between primary producers (phytoplankton) and higher trophic levels through energy transfer (Gokhale and Sharma 2022) and maintaining ecological balance by regulating phytoplankton populations by grazing and nutrient resupplying through excretion (see Hunt and Matveev 2005 for further details on these interactions). Studies employing this taxonomic group with regard to TiO_2 NPs are, therefore, crucial.

One study investigated the acute and chronic toxicity of TiO_2 NPs on the freshwater rotifer *Brachionus calyciflorus* at different temperatures (Dong et al. 2020). At 25°C, the 24 and 48-hour LC_{50} values were 117.14 and 60.11 mg L^{-1}, respectively. Exposure to TiO_2 NPs significantly affected various life history traits of *B. calyciflorus* at 15°C, 20°C, 25°C, and 30°C, decreasing life expectancy at birth, net reproductive rate, generation time, average lifespan, and/or intrinsic rate of population increase ($p < 0.05$). The toxicity of TiO_2 NPs to rotifers was enhanced at higher temperatures. An increase in the swimming linear speed of rotifers was observed in treatments with 200 µg L^{-1} TiO_2 NPs compared to the control group. Additionally, exposure to TiO_2 NPs concentrations ranging from 8 µg L^{-1} to 5 mg L^{-1} significantly increased superoxide dismutase activity. Glutathione content and catalase activity initially increased at 8 µg L^{-1} TiO_2 NPs exposure but significantly decreased in treatments at concentrations ranging from 40 µg L^{-1} to 5 mg L^{-1}. No significant changes in malondialdehyde contents among TiO_2 NPs treatments and the control group were noted. Overall, the study demonstrated that nTiO$_2$ exhibited high toxicity to rotifers, indicating significant environmental risks to aquatic ecosystems.

Another study extended TiO$_2$ NPs exposure times to 72 to 96 hours and investigated the toxicity of TiO$_2$ NPs suspensions with an initial mean diameter of approximately 100 nm on *Daphnia magna* (Dabrunz et al. 2011). The nominal concentrations of 3.8 mg L^{-1} (72-h EC50) and 0.73 mg/L (96-h EC50) resulted in toxicity. However, the TiO$_2$ NPs rapidly disappeared from the water phase, leading to lower toxicity levels of 0.24 mg L^{-1} (96-h EC50) based on measured concentrations. Furthermore, the toxicity of TiO$_2$ NPs (~ 100 nm) was compared to non-nanosized TiO$_2$ (~ 200 nm) prepared from the same stock suspension, and the findings indicate that the nanosized form was significantly more toxic. A mechanistic explanation for TiO$_2$ NPs toxicity in *D. magna* was proposed by the authors. Neonate *D. magna* ($\leq$ 6 hours) exposed to 2 mg L^{-1} TiO$_2$ NPs experienced a "biological surface coating" that disappeared within 36 hours, coinciding with successful molting in all exposed organisms. However, prolonged exposure up to 96 hours resulted in the reformation of the surface coating and a significantly reduced molting rate of only 10%, leading to 90% mortality. This suggests that D. magna coating with TiO$_2$ NPs combined with molting disruption, is a key factor in the observed toxicity. Given that the coating of aquatic organisms by manmade nanoparticles may occur widely in nature, this form of physical nanoparticle toxicity could have widespread negative impacts on environmental health.

In contrast, one study employing a novel approach combining traditional ecotoxicology methods with metal-specific recombinant biosensors to distinguish between the toxic effects of T nano and bulk TiO$_2$ and solubilized metal ions on *Daphnia magna* and *Thamnocephalus platyurus* verified that nano and bulk TiO$_2$ suspensions were non-toxic, even at an extremely high concentration of 20 g L^{-1} (Heinlaan et al. 2008). The authors, thus, indicate that metal oxide particles do not necessarily have to enter the cells to cause toxicity, and that the intimate contact between the crustacean gut environment and the particles seems to be more important, which may cause microenvironment changes in the vicinity of organism-particle contact area and either increase metal solubilization or generate extracellular ROS that may damage cell membranes.

Coral

Coral are marine invertebrates found in a variety of marine environments, primarily in shallow, warm waters, forming colonies of individuals, known as polyps, that secrete calcium carbonate skeletons (Goreau et al. 1979). Groups of coral form intricate structures known as coral reefs, built from the accumulated calcium carbonate skeletons of generations of coral polyps (Freiwald et al. 2004).

Corals have a symbiotic relationship with photosynthetic algae called zooxanthellae, which live within their tissues. The algae provide the corals with energy through photosynthesis, while the corals provide the algae with a protected environment and access to sunlight (Wooldridge 2010). When corals are subjected to prolonged stress, such as chemical contamination, they expel these symbiotic algae from their tissues, causing the coral to appear pale or "bleached". Several instances of coral bleaching events have been observed worldwide in the last decades, resulting in significant ecological effects, due to the unique and vital role of corals in marine ecosystems (Eakin et al. 2019).

Although many sunscreens are marketed as "reef-safe", as they do not contain, banned organic compounds such as such as BM-DBM, HS, or EHS (Miller et al. 2021), both ZnO and TiO_2 NP are added to sunscreens as inorganic UV filters. This has led to extremely high amounts of TiO_2 NP released from sunscreens in coastal zones due to increasing human populations and, consequently, activities, in these areas. In fact, about 496 million people live in coastal areas around the world in 2020, with estimates indicating a further increase of 143 million people by 2035 (Maul and Duedall 2019). This, in turn, directly affects a multitude of coastal species, including coral reefs.

Several laboratory studies have demonstrated coral bleaching in response to TiO_2 NP exposure. Jovanović and Guzmán (2014) for example, verified TiO_2 NP effects on the Caribbean reef-building coral (*Montastraea faveolata*). *Montastraea faveolata* specimens were and exposed for 17 days to TiO_2 NP (0.1 mg L^{-1} and 10 mg L^{-1}). TiO_2 NP exposure caused significant zooxanthellae expulsion in all coral colonies, without mortality. Induction of the gene for heat-shock protein 70 (HSP70) was observed during an early stage of exposure (day 2), indicating acute stress. However, there was no statistical difference in HSP70 expression on day 7 or 17, indicating possible coral acclimation and recovery from stress. No other genes were significantly upregulated. Inductively coupled plasma mass spectrometry analysis revealed that TiO_2 NP was predominantly trapped and stored within the posterior layer of the coral fragment (burrowing sponges, bacterial and fungal mats). The bioconcentration factor in the posterior layer was close to 600 after exposure to 10 mg L^{-1} of TiO_2 NP for 17 days. The transient increase in HSP70, expulsion of zooxanthellae, and bioaccumulation of TiO_2 NP in the microflora of the coral colony indicate the potential of such exposure to induce stress and possibly contribute to an overall decrease in coral populations. Another study assessed the effects of inorganic UV filters, including TiO_2 NP in two commercial formulations, Eusolex T2000 and Optisol, contained in sunscreen products on tropical stony corals (*Acropora* spp.), the predominant stony corals in reefs worldwide (Corinaldesi et al. 2018). Corals were exposed to 6.3 mg kg^{-1} TiO_2 NP for 48 h, leading to significantly higher zooxanthellae release compared to Eusolex at the initial time and 24-h timepoints and Optisol at 24 to 48 h exposure. However, no visible coral bleaching was evident compared to the controls.

Bleached coral exhibit reduced growth rates, decreased reproductive capacity, increased susceptibility to diseases and elevated mortality rates (Reef Resilience Network 2022). This, in turn, leads to a cascade of deleterious effects reef ecosystems, such as coral community composition changes which, in turn, affect the abundance and composition of many vertebrate and invertebrate species that directly depend on live coral for food, shelter, or recruitment habitats, ultimately resulting in genetic and species diversity declines in these already in fragile environments (Reef Resilience Network 2022).

It is also important to note that coral bleaching and mortality also lead to severe socioeconomic impacts, as they provide several significant ecosystem services, defined by the Millennium Ecosystem Assessment (Solomon 2023) as Provisioning services (comprising material or energy outputs from ecosystems, including food, raw materials, freshwater and medicinal resources); Regulating services (services that ecosystems provide by acting as regulators, such as Local climate and air quality, Carbon sequestration and storage, Moderation of extreme events, Waste-water treatment, Erosion prevention and maintenance of soil fertility, pollination and biological control); Supporting services (which underpin almost all other services, consisting of habitats for living organisms and maintenance of genetic diversity) and Cultural services (non-material benefits people obtain from contact with ecosystems, such as recreation, mental and physical health, tourism, aesthetic appreciation and inspiration for culture, art, and design and spiritual experience and sense of place).

Concerning coral bleaching and associated mortality, socio economic impacts include reduced shoreline protection services, as these organisms play a role as barriers against storms and large waves, aesthetic appeal, decreasing tourism activities and threatening the livelihoods of local communities, significant shifts in fish communities, in turn decreasing fish stocks, impacting food supplying and associated economic activities, and decreased sources of pharmaceutical compounds (Reef Resilience Network 2022).

Plants

Titanium, as other metals, is primarily taken up by plants through their root systems (Shahid et al. 2012). Once inside the plant, it is distributed throughout various plant tissues. Very contrasting findings have, however, been reported regarding TiO_2 in plants, with widely varying differential responses according to TiO_2 NP size, soil concentrations, shapes, doses, and exposure duration (Cox et al. 2016). Some studies have, for example, suggested potential beneficial Ti effects on plant growth under certain conditions, enhancing the tolerance of plants to abiotic stresses such as drought, salinity, and metal toxicity. However, the mechanisms by which titanium may confer these benefits are not yet fully understood. On the other hand, high Ti levels in plants can also have negative effects, interfering with the uptake and availability of essential nutrients, potentially leading to nutrient imbalances and adverse plant growth and development effects.

In one study, above- and below-ground growth, and root-tip cell mitosis of broad beans (*Vicia faba* L.) were assessed following exposure to TiO_2 at 25, 50 and 75 mg L^{-1} for 24 h (Thabet et al. 2019). The authors report decreased vigor index, reflecting shorter shoots at all exposure concentrations, but no effects on germination percentage and root length. Furthermore, no significant differences in the mitotic index were observed between the exposed groups and the control, although total chromosomal aberrations increased dose-independently, with breaks induced from 50 mg L^{-1}. In another assessment, *Allium cepa* meristematic cells were exposed to TiO_2 at 1000 mg L^{-1} (Santos Filho et al. 2019), resulting in slight seed germination and root growth inhibition, as well as severe cellular and DNA damage noted in a concentration-dependent manner following exposure to 10, 100, and 1000 mg L^{-1} TiO_2. Defense mechanisms were assessed at the highest tested concentration of 1000 mg L^{-1}, and nucleolar alterations and plant defence responses comprising increased lytic vacuoles and oil bodies were detected, indicating that, although TiO_2 NP can result in genotoxicity to *Allium cepa*, this species also displays defence mechanisms against this nanocompound. Another assessment aimed to evaluate TiO_2 NP effects *Lemna minor* (Li et al. 2013), reporting that these nanocompounds led to no adverse growth rate or chlorophyll a content effects, even at a high exposure concentration of 5 mg L^{-1} and extended exposure time of 14 days. Furthermore, although TiO_2 NP attached onto *L. minor* cell walls, no cellular uptake was observed. The authors draw attention to the fact that, although TiO_2 NP were not toxic to *L. minor*, the potential transfer of these compounds in aquatic food chains, such as attached to the plant leaves and other biological surfaces, must be considered in environmental risk assessments.

In another assessment, differential TiO_2 NPs uptakes and effects were noted when comparing wheat (*Triticum aestivum*) and rapeseed (*Brassica napus*) under hydroponic conditions (Larue et al. 2012). Wheat and rapeseed plantlets were subjected to 14 nm or 25 nm TiO_2 NP through root or leaf exposure. Titanium absorption was quantified using microparticle-induced x-ray emission in conjunction with Rutherford backscattering spectroscopy and distribution in roots and leaves was evaluated using micro x-ray fluorescence with synchrotron radiation. The results demonstrate that both TiO_2 NP sizes accumulated in the plantlets when exposed through the roots, with rapeseed exhibiting higher Ti content compared to wheat. The distribution of Ti in root cross sections varied depending on the agglomeration state of the nanoparticles. Additionally, the plantlets accumulated TiO_2 NP when exposed through the leaves. Furthermore, TiO_2 NP exposure led to increased root elongation but had no significant effect on germination, evapotranspiration, and plant biomass.

Another study aimed to assess the genotoxicity of TiO_2 NPs through classical genotoxic endpoints including the comet assay and DNA laddering technique in *Allium cepa* and *Nicotiana tabacum* (Ghosh et al. 2010). The comet assay and DNA laddering experiments confirmed the ability of TiO_2 NP to induce DNA damage in plant systems. In the case of Allium, micronuclei and chromosomal aberrations were observed, which were accompanied by a reduction in root growth. Additionally, increased malondialdehyde (MDA) concentrations were observed at a treatment dose of 4 mmol L^{-1} of TiO_2 NP in *Allium cepa*, suggesting that lipid peroxidation may be involved as one of the mechanisms leading to DNA damage in plants following TiO_2 NP exposure.

Non-toxic TiO_2 effects, on the other hand, have been reported concerning the germination and root elongation of seed and seedlings in *Brassica campestris* ssp. *napus* var. *nippo-oleifera* Makina (oilseed rape), *Lactuca sativa* L. (lettuce), and *Phaseolus vulgaris* var. *humilis* (kidney bean) in both *in vitro* and *in situ* assessments with no effect on total antioxidant capacity and superoxide dismutase activity or chlorophyll content, even though TiO_2 NP absorption was detected (Song et al. 2013).

45

Finally, and very interestingly, positive TiO_2 NP effects has also been noted in plants. One study assessed the effects of TiO_2 NPs on the agronomic traits of Moldavian balm (*Dracocephalum moldavica* L.) in a greenhouse experiment. The plants were subjected to different salinity levels (0, 50, and 100 mmol L^{-1} NaCl), and TiO_2 NPs were applied at 0, 50, 100, and 200 mg L^{-1}. The results indicated that all agronomic traits were negatively impacted by salinity levels, but the application of 100 mg L^{-1} TiO_2 NPs mitigated the noted adverse effects, improving all agronomic traits and increasing the activity of antioxidant enzymes compared to plants grown under salinity without TiO_2 NPs treatment. Furthermore, the application of TiO_2 NPs significantly reduced the concentration of hydrogen peroxide, a known oxidant. The highest content of essential oil (1.19%) was also observed in plants treated with 100 mg L^{-1} TiO_2 NPs under control conditions, with a comprehensive GC/MS analysis of the essential oils revealing that geranial, z-citral, geranyl acetate, and geraniol were the predominant components. The highest geranial, geraniol, and z-citral levels were found in plants treated with 100 mg mg L^{-1} TiO_2 NPs under control conditions. In conclusion, the application of 100 mg L^{-1} TiO_2 NPs demonstrated a significant ameliorative effect on the salinity-induced effects in Moldavian balm.

Mollusks

Mollusks play a significant role in various ecosystems and exhibit considerable importance both ecologically and economically, the latter in aquaculture and fisheries industries, comprising a valuable resource in providing a sustainable source of seafood and supporting local economies (Bostock et al. 2016).

Due to several specific features, including their sessile and filter-feeder nature, wide distribution, easy sampling, high abundance, low mobility, and ecological and economic importance, as well as sensitivity to pollutants, bivalve mollusks have been noted as one of the best sentinel organisms concerning several classes of contaminants (Laitano and Fernández-Gimnénez 2016), routinely used to assess aquatic ecosystem health and quality. They are, in fact, employed in the Mussel Watch Program, a long-term environmental monitoring program that focuses on the assessment of pollution and contamination levels in coastal and estuarine environments (Soto et al. 2000). This program involves collecting and analyzing tissue samples from bivalve mollusks, specifically mussels, at various coastal locations. Initially established by the National Oceanic and Atmospheric Administration (NOAA) in the early 1980s in the United States, this program has been implemented in numerous countries worldwide, with the main aim to monitor and track the presence and trends of chemical contaminants, including metals, organic pollutants, and emerging contaminants, in coastal ecosystems over time. The Mussel Watch Program plays a vital role in understanding contaminant distribution, trends, and potential ecological risks in coastal ecosystems (Guitart et al. 2012), aiding in identifying pollution sources, assessing the effectiveness of pollution control measures, and guiding decision-making processes related to environmental management and policy. Due to its long-term nature, the program enables the assessment of changes in contamination levels over time, providing valuable insights into the efficacy of environmental regulations and the impact of pollution reduction efforts.

In this regard, many studies concerning TiO_2 NPs have been conducted with this group to assess different physiological effects.

In one assessment, the neurotoxic effects of a 96-h acute TiO_2 NPs exposure on the benthic marine bivalve blood clam, *Tegillarca granosa* were assessed by evaluating three major neurotransmitters (DA, GABA, and ACh), acetylcholinesterase activity and the expression of neurotransmitter-related genes (Guan et al. 2018). The three neurotransmitters (DA, GABA, and ACh) were significantly increased when exposed to 1 mg L^{-1} TiO_2 NPs for DA and 10 mg L^{-1} TiO_2 NPs for ACh and GABA. Furthermore, TiO_2 NPs led to lower AChE activity and down-regulation of the expression of genes encoding modulatory enzymes (AChE, GABAT, and MAO) and receptors (mAChR3, GABAD, and DRD3). Thus, the authors conclude that TiO_2 NP exposure results in significant neurotoxic effects in *T. granosa*, indicating the potential for the disruption on several physiological processes.

Concerning TiO_2 NP mechanism of action in bivalves, Guan et al. (2019) conducted a acute (96 h) TiO_2 NP exposure and recovery trials in the blood clam, *Tegillarca granos, reporting suppression on* phagocytosis rate, cell viability, and intracellular Ca^{2+} haemocyte concentrations and significantly increased intracellular ROS concentrations, as well as the downregulation of Caspase-3, Caspase-6, apoptosis regulator Bcl-2, Bcl-2-associated X, calmodulin kinase II, and calmodulin kinase II, all of which were partially mitigated by the addition of exogenous Ca^{2+}. indicating that Ca^{2+} signalling could be one of the key pathways through which TiO_2 NPs attacks phagocytosis, leading to immunotoxicity. Another assessment in this regard aimed at investigating the mechanisms of action of TiO_2 NPs in bivalve molluscs carried out *in vitro* exposure of clam (*Ruditapes philippinarum*) haemocytes to TiO_2 NPs to evaluate effects on phagocytic activity and NPs internalisation into haemocytes. The study focused on evaluating the effects of n TiO_2 NPs on haemocyte phagocytic activity through two different experiments: one with pre-treatment of haemocytes and one without. Cells were exposed to P_{25} TiO_2 NPs at 0, 1, and 10 µg mL^{-1}. Additionally, transmission electron microscopy (TEM) was also employed to investigate the interaction between TiO_2 NPs and clam haemocytes. The results in both experimental setups revealed that TiO_2 NPs significantly reduced the phagocytic index compared to the control group, indicating that NPs display the ability to disrupt cell functions. This hypothesis was further supported by the TEM analysis, which demonstrated the interaction of TiO_2 NPs with cell membranes and their entry into haemocyte cytoplasm and vacuoles after 60 minutes of exposure.

In another assessment, Lu et al. (2019) attempted to establish a baseline for total Ti in marine bivalves, namely oysters, mussels, and clams, from China and other regions around the globe through a meta-analysis and modeling approach, using probability frequency distributions. The findings indicate that baseline Ti concentrations were similar among the investigated bivalve species and that concentrations of this element in China are very similar to those reported for other areas worldwide.

Interestingly, some studies have also evaluated TiO_2 NPs toxicity under probable climate change scenarios in molluscs. For example, one study, the effects of TiO_2 NPs on digestive enzyme activities were assessed in the marine mussel *Mytilus coruscus* in an ocean acidification context (Kong et al. 2019), due to the significant concerns of this climate change effect and its worldwide increase. The mussels were exposed to combined treatments at 0, 2.5 and 10 mg L^{-1} TiO_2 NP concentrations and pH values of 7.3 and 8.1 for 14 days, and then recovered under ambient conditions at pH 8.1 with no TiO_2 NPs for 7 days. Samples were taken on the 1st, 3rd, 7th, 14th, and 21st day, the digestive enzymes, including amylase, pepsin, trypsin, lipase, and lysozyme, were investigated. Separately, TiO_2 NPs and low pH led to negative amylase, pepsin, trypsin, and lipase effects, while both combined led an increase in lysozyme activity. The authors indicate that TiO_2 NPs was a more significant deleterious factor than pH, and that a recovery period of 7 days was not sufficient for the assessed enzymes to fully recover initial pre-experimental activities. Other assessments in ocean acidification scenarios corroborate these findings, reporting that ocean acidification seems to stimulate TiO_2 NPs accumulation in bivalves and lead to significantly deleterious physiological effects. For example, 1.34- and 1.16-fold higher TiO_2 NPs accumulation was noted in *T. granosa*, *M. meretrix*, and *C. sinensis* raised at low pH values (7.4 and 7.8), compared to clams at pH 8.1 (Shi et al. 2019), while higher Malondialdehyde (MDA) levels and decreased superoxide dismutase (SOD) activity and reduced glutathione (GSH) concentrations were reported in *M. coruscus* exposed to 2.5 and 10 mg L^{-1} TiO_2 NPs and pH 7.3 and 8.1 for 14 days, with a 7-day recovery period also noted as not enough for recovery to baseline levels (Huang et al. 2018).

Higher crustaceans

Titanium assessments in crustaceans like shrimp, crabs, and lobsters, are still scarce, even though these animals are a staple in the diet of many human populations, comprising a source of high-quality protein, vitamins, minerals, and omega-3 fatty acids, contributing to a healthy and balanced diet, as well as important components of trophic webs. These animals also have significant economic value as commercial fisheries and aquaculture targets, both in marine and freshwater environments (Susanto 2021).

In one study, total Ti was quantified in the muscle of two crustacean groups, swimming crabs (*Callinectes sapidus* and *Achelous spinimanus*) and shrimp (*Farfantepenaeus paulensis* and *Litopenaeus schmitti*) from one of Brazil's most important, albeit highly polluted estuary, located in Rio de Janeiro, in the country's southeastern region (de Almeida Rodrigues et al. 2022). The authors reported that the determined Ti concentrations were higher than in other assessments worldwide, and that animal length and weight, as well as the abiotic factors depth, transparency, dissolved oxygen, and salinity, significantly influence crustacean Ti concentrations. Furthermore, a human health risk assessment was also performed, and potentially significant risks were noted after calculating a simulated exposure to TiO_2.

In one study conducted in India, total Ti was determined in forty-five assorted sea food samples collected every month over a period of 14 months from three different seafood landing centres (Supriya et al. 2020). Values in crabs ranged from 1.00 to 1.50 mg kg^{-1} w.w. and in prawns, from 0.57 to 0.82 mg kg^{-1} w.w., indicating inter-group differences. The authors attribute these relatively high Ti values to high levels of this element in sediments in different coastal areas near the sampling sites, as high as 4100–20,000 mg L^{-1}.

In another assessment, Wang et al. (2023) investigated the toxicity of TiO_2 NP exposure and the underlying mechanisms on spermatogenesis and adhesion junctions in male Chinese mitten crabs *Eriocheir sinensis* testes following exposure to 3 nm and 25 nm TiO_2 NP at a 30 mg kg^{-1} body weight dose. The results indicate that both TiO_2 NP sizes induced apoptosis and damaged the integrity of the haemolymph-testis-barrier, a structure similar to the blood-testis-barrier, as well as the structure of the seminiferous tubule. Furthermore, the 3 nm TiO_2 NP caused more severe spermatogenesis dysfunction compared to the 25 nm TiO_2 NP. Furthermore, TiO_2 NP exposure affected the expression patterns of adherens junctions (α-catenin and β-catenin) and caused tubulin testis disorganization and genefated reactive oxygen species and an imbalance in the mTORC1-mTORC2 pathway. Specifically, mTORC1, rps6, and Akt levels increased, while mTORC2 activity remained unchanged. The findings suggest that the mTORC1-mTORC2 imbalance induced by TiO_2 NP is involved in the disruption of adherens junctions and the haemolymph-testis-barrier, ultimately leading to spermatogenesis dysfunction in *E. sinensis*.

Fish

Fish, as the other taxonomic groups mentioned above, play vital roles in aquatic ecosystems, as key components of food webs, occupying various trophic levels and, thus, contributing to energy transfer and nutrient cycling within aquatic ecosystems. They also provide substantial socioeconomic benefits globally, contributing to employment, income generation, and food security for many coastal communities and nations (McIntyre et al. 2016). Fish are, in fact, one of the most important sources of animal protein, essential fatty acids (such as omega-3 fatty acids), vitamins (including vitamin D and B vitamins), and minerals (such as iodine, selenium, and zinc) in many human diets (Mishra 2020).

Teleosts

Many studies have been conducted on model fish species, such as zebrafish (*Danio rerio*). In this regard, TiO_2 NPs were shown to be absorbed and systemically transported in these fish, also able of traversing the blood–heart barrier and accumulate in the heart (Chen et al. 2011), leading to several negative effects, including sparse cardiac muscle fibres, inflammatory responses, cell necrosis and cardiac biochemical imbalances (Chen et al. 2011). In another assessment, sub-lethal TiO_2 NP effects on the physiology and reproduction of zebrafish were evaluated following exposure for 14 days at (0.1 or 1.0 mg L^{-1}) (Ramsden et al. 2013). No change in erythrocyte counts were observed, while a two-fold decline in leukocyte counts was noted. No changes in Na^+K^+-ATPase activity in brain, gill or liver tissues were detected, although total

51

glutathione (GSH) levels in brain, gill and liver tissues were higher in exposed fish. No significant effects at the histological level were detected. At the end of the 14-d exposure adult zebrafish, however, were able to reproduce, although with a lower number of viable embryos. A further evaluation aimed to verify TiO_2 nanoparticle-induced neurotoxicity in zebrafish, assessing morphological changes, neurochemical alterations, and the expressions of memory behaviour-related genes in brains following to 5, 10, 20, and 40 μg L^{-1} TiO_2 NPs exposure for 45 days (Sheng et al. 2016). The findings indicated significantly reduced spatial recognition memory and norepinephrine, dopamine, and 5-hydroxytryptamine levels, but markedly high NO levels, as well as over proliferation of glial cells and neuron apoptosis, after low dose exposures. Low doses also significantly activated *C-fos, C-jun,* and *BDNF* genes, and suppressed expressions of *p38, NGF, CREB, NR1, NR2ab,* and *GluR2* genes. The findings, thus, indicate significant brain damages in zebrafish.

Several assessments have been conducted in other, non-model, teleosts. One of these, for example, evaluated the genotoxicity, potential cytotoxicity and cell uptake of titanium dioxide nanoparticles in the marine fish *Trachinotus carolinus*. The fish were administered two different TiO_2 NPs doses (1.5 μg and 3.0 μg g^{-1}) intraperitoneally. Blood samples were collected at 24, 48, and 72 hours following the injections to assess erythrocyte viability using the Trypan Blue exclusion test, comet assay (pH > 13), micronucleus (MN) assay, and other erythrocyte nuclear abnormalities (ENA) examination. After 72 hours, the potential uptake of TiO_2 NPs by cells was investigated in fish that received the higher dose using transmission electron microscopy (TEM). The findings indicated that TiO_2 NPs exhibited genotoxic and potentially cytotoxic effects on this species, resulting in DNA damage, triggering the formation of micronuclei and other erythrocyte nuclear abnormalities, and reduced erythrocyte viability. A transmission electron microscopy analysis revealed that TiO_2 NPs were primarily taken up by cells in the kidney, liver, gills, and to a lesser extent, in muscle tissue.

As noted, teleost assessments regarding TiO_2 NP toxicity effects in general are relatively abundant in the literature. Studies on neotropical teleosts, however, are not as widespread, although some in this regard are available. For example, one study evaluated TiO_2 NP effects in a Neotropical detritivorous fish, *Prochilodus lineatus* following exposure to 0, 1, 5, 10, and 50 mg L^{-1} TiO_2 NP for 48 hours and for 14 days (Carmo et al. 2019). The findings indicated no red blood cell damage, an acute decrease in white blood cell concentrations and increased monocyte concentrations in the acute exposure, while the chronic exposure resulted in reduced red and white blood cell concentrations and lymphocytes and increased mean cell volume and hemoglobin, which the authors report as indicative of significant immune system alterations and increased energy expenditure. The acute exposure resulted in increased ROS and GSH in liver, while the chronic exposure resulted in decreased superoxide dismutase activity and increased glutathione-S-transferase (GST) activity and GSH content. Furthermore, acetylcholinesterase was also decreased in muscle following acute TiO_2 NP exposure, indicating neurotoxic potential. Another study aimed to assess the interaction between TiO_2 NPs and incident light in the neotropical *Piaractus mesopotamicus* exposed for 96 h to 0, 1, 10, and 100 mg L^{-1} TiO_2 NP, under visible light, and visible light with ultraviolet (UV) light (Clemente et al. 2013). No mortality was observed under any of the test conditions, while several sublethal effects influenced by illumination condition were observed, such as acid phosphatase activity inhibition under both illumination conditions following exposure to 100 mg L^{-1}, increased metallothionein levels in fish exposed to 1 mg L^{-1} TiO_2 NP under visible light, reduced protein carbonylation in groups exposed to 1 and 10 mg L^1 under UV light and higher nucleus alterations in erythrocytes in fish exposed to 10 mg L^1.

Another recent assessment (Oliveira et al. 2023) investigated the effects of titanium dioxide TiO_2 NP on sperm motility and spermatozoa population structure using a subpopulation approach, where seabream sperm samples from mature males were exposed to increasing concentrations of titanium dioxide TiO_2 NP for 1 hour. The concentrations chosen included both realistic and supra-environmental values. After the *ex vivo* exposure, sperm motility parameters were analysed using computer-assisted sperm analysis, and sperm subpopulations were identified using a two-step cluster analysis. The results indicate a significant reduction in total motility when exposed to the two highest concentrations of TiO_2 NP, while curvilinear and straight-line velocities were unaffected. The exposure to both TiO_2 and Ag NPs also affected sperm subpopulations. Higher NP levels led to a decrease in the percentage of fast sperm subpopulations and an increase in slow sperm subpopulations. Overall, the findings suggest a reprotoxic effect of TiO_2 NP, but only at supra-environmental concentrations.

Elasmobranchs

Interestingly, only one study has reported Ti concentrations in elasmobranchs to date, namely an assessment concerning total and intracellular metal concentrations and metallothionein metal detoxification in blue sharks, *Prionace glauca* L. from the Western North Atlantic Ocean (Hauser-Davis et al. 2021). Total mean Ti contents were determined as 19.69 ± 17.85 mg kg^{-1} wet weight in muscle and 2.31 ± 0.98 in liver, while thermostable, MT-bound, Ti content was reported as 1.063 ± 0.296 mg g^{-1} wet weight in liver and below the LOD in muscle. That study reported metallothionein Ti detoxification for the first time in elasmobranch liver, which was not noted in muscle.

As several shark and ray species present coastal habits during at least one life stage, they are probably highly exposed to this metal, although the notable lack of assessments in elasmobranchs has resulted in a severe knowledge gap concerning Ti levels and potentially harmful effects in this group. Thus, assessments in this regard are paramount for this taxon, especially as many elasmobranchs (391 species, or 32.6%) are threatened with extinction, mainly due to overfishing, although three other major threats have been recently described, namely loss and degradation of habitat, climate change and pollution (Dulvy et al. 2021).

Amphibians

Amphibians, which include frogs, toads, salamanders, and newts, playing a crucial role in ecosystems in ecological balance, contributing to energy transfer, particularly tadpoles and larvae, and nutrient cycling within aquatic ecosystems and influencing the overall ecological balance of their habitats. Their excretions, such as nitrogen and phosphorus, can also fertilize the surrounding land/ aquatic ecosystems.

Amphibian TiO_2 NP toxicity assessments are still uncommon, and most have been conducted in the model African clawed frog, *Xenopus laevis*. One study to this end evaluated the effects of a 96 h exposure with daily solution exchanges for TiO_2 NPs, among other nanomaterials, employing the Frog Embryo Teratogenesis Assay Xenopus (FETAX) protocol (Nations et al. 2011). The findings indicated no increased mortality in static renewal exposures containing up to 1000 mg L^{-1} TiO_2 NPs, although gastrointestinal and spinal developmental abnormalities were observed. In another assessment, viability and growth of *Xenopus laevis* were evaluated following exposure to eight TiO_2 NPs concentrations in the presence of either white light or UVA. The findings indicate that, increasing TiO_2 concentration decreased X. *laevis* survival regardless of light exposure, also significantly affecting tadpole growth and delaying developmental stages (Zhang et al. 2012).

In one of the scarce studies available in the literature on other species, TiO_2 NPs effects in Neotropical *Dendropsophus minutus* tadpoles exposed to three different concentrations (0.1 mg·L^{-1}, 1.0 mg·L^{-1}, and 10 mg·L^{-1}) were evaluated (do Amaral et al. 2022). Significant DNA damage was noted, as well as decreased total size, body length, width, and height of the tail musculature. In a behavioural test, tadpoles exposed to TiO_2 NPs presented reduced mobility and less aggregation. Another study assessed TiO_2 NPs on cultured *Rana catesbeiana* tailfin tissue (Hammond et al. 2013). xcultured tailfin assay was used to examine exposure effects of 8–800 ng L^{-1} of three types of ~ 20 nm TiO_2 NPs (P25, M212, M262). Real-time quantitative polymerase chain reaction indicated no significant effect on TH-responsive transcripts, some significant effects on stress-related transcripts following exposure to micron-sized TiO_2, P25, and M212 and no effect after M262 exposure. The authors conclude that adverse amphibian tissue effects through TH-signalling, or cellular stress seems low for the tested TiO_2 NPs.

Another assessment evaluated the effects of TiO_2 NPs on the survival and growth of *Sclerophrys arabica* tadpoles in a two-level trophic system, exposed directly in water, indirectly through their food source (decomposing leaves), or a combination of both (Al Mahrouqi et al. 2018). Tadpoles fed TiO_2 NP-treated leaves grew larger than controls, which the authors postulate as being due to more palatable or nutritious leaf material for the tadpoles due to TiO_2 NP action. This was observed only in natural light conditions, as TiO_2 NPs can also directly damage leaf structure, potentially making the leaves more digestible for tadpoles, However, toxicity was noted via direct water-borne exposure but enhanced through consumption of TiO_2 NP exposed leaves.

Reptiles

Reptiles represent a diverse group of vertebrates, which include snakes, lizards, turtles, and crocodilians. These animals inhabit a range of terrestrial, freshwater, and marine habitats and hold ecological and conservation importance, playing significant roles in ecosystems, such as seed dispersal, contributing to vegetation diversity and ecosystem stability. In addition, some species, mainly turtles and crocodilians play important roles in folklore, traditional practices, and spiritual belief, while also being commercially valuable for their skin, meat, and other body parts, contributing to local economies (da Nóbrega Alves et al. 2008, Alexander et al. 2017). However, many members of this group are extremely threatened by anthropogenic activities. The existing body of ecotoxicological literature, however, is recognized for its insufficient representation of reptiles (Grillitsch and Schiesari 2010). Interestingly, turtles are the main studied group regarding Ti contamination, while scarce studies on lizards and crocodilians are available, and no assessments on snakes were found. It is interesting to note in this regard that several studies assessed not only the animals themselves, but also eggs. The use of reptilian eggs is extremely interesting in ecotoxicological assessments, as it is a non-invasive method and provides information on contaminant levels in both females and the developing embryos (Klein et al. 2012, Bouwman et al. 2014, du Preez et al. 2018). Furthermore, besides the contents, eggshells also have toxicological implications, as pollutants are known to distribute between the shell and its contents (Kleinow et al. 1999).

One study evaluated Ti, along with other metals, in stranded dead Olive Ridley turtles (n = 17) from *La Escobilla* beach, in the state of Oaxaca on the Mexican Pacific, by inductively coupled plasma optical emission spectrometry (ICP-OES) (Cortés-Gómez et al. 2018a). This elements was detected in yolk (0.03 ± 0.02 µg g^{-1} wet weight), albumin (0.02 ± 0.01 µg g^{-1} wet weight), liver (0.6 ± 1.0 µg g^{-1} wet weight), kidney (1.2 ± 1.1 µg g^{-1} wet weight), muscle (2.2 ± 3.1 µg g^{-1} wet weight), brain (9.2 ± 12 µg g^{-1} wet weight), and bone (9.6 ± 7.3 µg g^{-1} wet weight), blood (7.7 ± 24 µg g^{-1} wet weight), and not detected in eggshells. Ti displayed a significant relationship with a developmental instability index (DIx) used to determine turtle carapace asymmetry in blood (positive association) and albumin (negative association). The authors report that bone and brain presented the highest Ti concentrations (9.6 and 9.1 µg g^{-1} respectively) and that more symmetric turtles excreted higher Ti levels through egg yolk and albumin, than more asymmetric individuals, possibly due to the detoxification role that eggs play in some turtle species (Nagle et al. 2001, Guirlet et al. 2008), although further studies are required in this regard.

In another study, Ti levels were determined in the plasma of Green sea turtle from two different coasts in Japan, one suburban (n = 8) and one coastal (n = 57) by Particle induced X-ray emission (Tsukano et al. 2017). Higher levels noted in the plasma of individuals from the suburban coast, which the authors associated to the influence of residential and industrial wastewater, mainly due to TiO$_2$ NP use in household products.

Another assessment determined Ti concentrations in the serum of 44 healthy nesting female Olive Ridley turtles (*Lepidochelys olivacea*) from *La Escobilla* beach, in the state of Oaxaca in the Eastern Pacific (Southeast Mexico) by ICP-OES and evaluated associations with *p-nitrophenyl acetate esterase activity* (EA) and cortisol (Cortés-Gómez et al. 2018b). The findings indicated Ti concentrations of 0.99 ± 1.84 (6.39 ± 11.87) dry weight and significant correlation between Ti and EA ($r = -0.37$, $p < 0.01$), suggesting that turtles chronically exposed to this element suffer high esterase consumption and prolonged cortisol increases. The authors indicate that the negative relationship with Ti reinforces the results of other researchers in humans regarding possible EA inhibition by metals.

Another report assessed the serum of 100 nesting Olive Ridley turtles (*Lepidochelys olivacea*) from *La Escobilla* beach, in the state of Oaxaca in the Eastern Pacific (Southeast Mexico) by ICP-OES (Cortés-Gómez et al. 2018c). Ti levels were 0.03 ± 0.08 µg g^{-1} (ranging from 0.01 to 0.61 µg g^{-1}) and four positive relationships were noted, with creatinine, urea, cholesterol, and cortisol. The authors postulate that cortisol could act as an anti-inflammatory corticosteroid against inflammation caused by Ti (Rhen and Cidlowski 2005).

Suzuki et al. (2012) analyzed the plasma of captive and wild hawksbill (*Eretmochelys imbricate*) Green sea turtles (*Chelonia mydas*) and Loggerhead sea turtles by Particle-Induced X-ray Emission Analysis, reporting the following values for captive animals: Hawksbill sea turtles (n = 25) – 0.303 ± 0.226 µg mL^{-1}, Green sea turtles (5) – 0.234 ± 0.266 µg mL^{-1} and Loggerhead sea turtles (3) – 0.137 ± 0.129 µg mL^{-1} and wild animals: Hawksbill sea turtles (n = 6) – 0.307 ± 0.179 µg mL^{-1}, Green sea turtles (n = 9) – 0.444 ± 0.353 µg mL^{-1} and Loggerhead sea turtles (n = 9) – 0.128 ± 0.129 µg mL^{-1}.

Concerning lizards, to the best of our knowledge, only one study has been conducted concerning Ti contamination, where McIntyre and Whiting (2012) assessed Ti concentrations, among other metals, in tissue and blood samples from adult Giant Sungazer Lizards (*Smaug giganteus*) from mining areas in South Africa. The study area encompasses a natural saltpan used for over 40 years for the authorized discharge and evaporation of gold mine process water, as well as, to a lesser extent, purified sewage effluent and is contaminated to several elements associated to acid mine drainage and the discharge of mine process water, including Ti. However, Four Ti concentrations were lower than minimum limits of detection for most of the analyzed samples. Of the samples above the limit, tail tissue values were determined as 21.3 mg kg^{-1} d.m. and 26.1 mg kg^{-1} d.m. at two mining sites and 16.7 mg kg^{-1} d.m. and 28.7 mg kg^{-1} d.m. at two control sites. It is important to note that, although the the primary metal uptake route in reptiles is expected to be dietary, the authors note that sungazer lizards are burrowing creatures, and that possibility of direct uptake from contaminated soils cannot be ruled out, as soil ingestion has been demonstrated as a potentially significant pathway for contaminant uptake in lizards (Rich and Talent 2009).

Similarly, only one assessment was found on crocodilians, where du Preez et al. (2018) determined total Ti concentrations alongside other metals in Nile crocodile (*Crocodylus nyloticus*) eggs from three different locations in the Kruger National Park in South Africa and compared the values to those in eggs obtained from a crocodile farm used as reference. Egg contents (yolk and albumin together) were determined separately from the shells. Egg contents ranged from 6.8 to 6.9 mg kg^{-1} dry mass (d.m.) at the Kruger National Park and as 7.0 mg kg^{-1} d.m. at the crocodile farm, while shell values were significantly lower, ranging from 0.19 to 0.69 mg kg^{-1} d.m. at the Kruger National Park and were determined as 0.29 mg kg^{-1} d.m. at the crocodile farm. No discussions, however, on Ti toxicity or potential sources were conducted in that study.

Birds

Birds also play a crucial role in ecosystems, providing numerous benefits to the environment and human societies, including but not limited to pollination, seed dispersal, pest control and nutrient cycling, as well as in ecotourism and education and ecosystem stability.

Very few studies have assessed Ti concentrations and reported on their toxic effects in birds. One review specifically focusing on seabirds (Hauser-Davis et al. 2020) reported only five studies concerning studies on Ti concentrations in this group, ranging from 0.35 to 6.23 mg kg^{-1} in liver, 1.85 to 3.78 mg kg^{-1} in kidneys and from 0.1 to 17 mg kg^{-1} in feathers, presenting significant interspecies variations. This, thus, demonstrates a significant knowledge gap concerning Ti in seabirds and indicates the urgent need to establish baseline data for this element in this group.

Concerning other groups of birds, one study aimed to detect genetic TiO_2 NP effects by PCR in the testes of 10 to 50-day-old quail treated orally with these compounds (Al-Jomily and Al-Sultan 2022). A significant difference was noted in the 35 mg kg^{-1} TiO_2 NP exposure group compared to the control, indicating potentially negative reproductive outcomes. Another study assessed the Histopathological effects of TiO_2 NP on the liver of Japanese quail *Coturnix coturnix japonica* (Ahmed and Taha 2022). A control group was dosed with distilled water for four continuous days, and two experimental groups were dosed with TiO_2 NP at 20 and 40 mg/kg, respectively. Four, fourteen, thirty and sixty days after the experiment began, the birds were sacrificed. The results indicated the development of several histological lesions in the liver of birds of the two experimental groups, to varying degrees, at four, fourteen, thirty and sixty days of exposure. The most prominent tissue lesions in the second experimental group, comprising necrosis, hemorrhage, vacuolation, congestion and ballooning swelling, as well as inflammatory cell infiltration, while the third experimental group, besides presenting histopathological lesions similar to the second group, also displayed sinuses dilatation, Kupffer cells hypertrophy, hepatocyte enlargement, and necrosis of the walls of blood vessels and bile ducts.

In another study, some vitamin E (VE) effects on Japanese quail performance, eggs and blood parameters during short-term contamination with high levels of Ag-coated TiO_2 NP were evaluated (Firouzifard et al. 2016). A total of ninety-six Japanese quails at 20 weeks of age were randomly divided into six treatment groups, each with four replicates consisting of four birds (three females and one male) per cage. The treatments included two levels of VE supplementation (0 and 200 mg kg^{-1} diet) and three levels of Ag-coated TiO_2 NP (0, 500, and 1000 mg kg^{-1} Diet). The results indicated that supplementation with Ag-coated TiO_2 NPs at the specified levels led to a significant increase in egg production, although no significant effects were observed on other performance parameters and egg quality parameters. Additionally, the plasma parameters were not influenced by Ag-coated TiO_2 NP. Although the interaction between Ag-coated TiO_2 NPs and VE was not statistically significant, the plasma levels of triglycerides and uric acid showed a tendency to decrease and increase, respectively in birds fed VE. Thus, the authors conclude that exposure of Japanese quails to Ag-coated TiO_2 NPs did not result in any toxic effects on their performance, egg quality, and stress-related plasma parameters.

Another assessment aimed to investigate the effects of incorporating a mixed powder of *Anacardium occidentale* leaves (60%), *Psidium guajava* leaves (20%), and *Morinda citrifolia* leaves (20%) on the growth performance, nutrient digestibility, and immunoglobulin concentration in broiler chickens, analysing Ti as a contaminant (Aroche et al. 2018). A total of 80 one-day-old Ross 308 broiler chickens were randomly divided into two dietary groups from days 1 to 21 of age. The treatments included a control diet (T0) and a diet supplemented with 0.5% of the mixed powder of medicinal plant leaves (T1). The inclusion of herbs resulted in a decrease in feed intake (FI) and feed conversion ratio (FCR) during the second week and throughout the entire study period (1–21 days) compared to T0, with Ti levels remaining the same at T0 and T1, at 0.30 mg kg^{-1}. No significant effect on body weight (BW) and average gain (AG) was noted, and the mixed powder did not affect nutrient digestibility and IgA concentrations but did lead to increased IgG concentrations. Based on the results, the authors recommend the inclusion of the mixed powder of medicinal plant leaves in the diet as an alternative to achieve satisfactory performance in broiler chickens. However, no Ti toxicity assessments were evaluated, and no discussion on this was conducted in that study.

In a similar study, however, supplemental protease with a phytase and xylanase and cereal grain source with reduced amino acid concentrations containing 5.0 g kg^{-1} TiO_2 was shown to affect broiler nutrient digestibility and performance (Ingelmann et al. 2018). At 14 days of age, a significant interaction between cereal grain and protease, which influenced starch digestibility in the jejunum and ileum. Protease improved digestibility in birds fed wheat-based diets but had no significant effect in those fed maize-based diets. The cereal grain source significantly affected nitrogen digestibility and digestible energy with birds fed wheat-based diets showing higher digestibility compared to those fed maize-based diets. At 28 days of age, birds fed wheat-based diets exhibited higher nitrogen digestibility in the jejunum and ileum, while protease supplementation reduced ileal nitrogen digestion. A significant interaction was noted between cereal grain and protease for starch digestion in the ileum, affecting birds fed wheat-based diets but not those fed maize-based diets. Wheat-based diets led to higher DE compared to maize-based diets in both the jejunum and ileum. From 15 to 35 days of age, cereal grain source significantly influenced performance, with broilers consuming maize-based diets exhibiting better performance than those fed wheat-based diets. The authors, however, discuss the data only in terms of digestibility, with no comments on the presence or effects of TiO_2.

Concerning birds, TiO_2 NPs have been evaluated as litter treatment agent concerning litter bacteriology, haematology, ammonia emission, and ammonia-related lesions in broiler production by Adamu et al. (2015). Different TiO_2 NPs concentrations (0 g m^{-2}, 30 g m^{-2}, 60 g m^{-2}, and 120 g m^{-2}) were applied to a deep litter system using wood shavings at a depth of 6 cm, with 16 birds per square meter. Litter samples were collected weekly to measure ammonia levels, while tissue samples were collected at the seventh week for histology. Blood samples were also obtained through cardiac venipuncture for haematology and blood chemistry analysis. The results revealed that red blood cell count (RBC) and white blood cell count (WBC) were significantly elevated in the TiO_2-treated litter at 120 g/m^2. Mean corpuscular haemoglobin (MCH) and mean corpuscular volume (MCV) were also significantly increased in the 120 g/m^2 TiO_2 litter treatment, as well as serum sodium, potassium, globulin, and conjugated bilirubin levels. Surprisingly, no significant difference in ammonia levels were noted, although the 60 g/m^2 TiO_2 litter treatment retained higher ammonia levels. The authors conclude that TiO_2 demonstrated beneficial effects as a litter treatment agent, reducing ammonia-associated lesions due to improved nitrogen retention, particularly when applied at 60 g/m^2.

Marine mammals

Marine mammals, which include whales, dolphins, seals, sea lions, manatees, and polar bears, play a vital role in marine ecosystems and hold significant ecological and conservation importance, contributing towards ecological balance by regulating prey species, nutrient cycling, carbon sequestration, particularly whales, which transport substantial amounts of carbon from the surface to deep ocean waters through their feeding behavior, aiding in mitigatin g climate change by removing carbon from the atmosphere and storing it in the deep ocean, as well as in tourism and recreational ctivities such as whale watching and dolphin encounters, which contribute to local economies and provide opportunities for education and conservation awareness.

Some studies concerning Ti evaluations in marine mammals are available, although, as for some of the other groups discussed herein, not many.

In one assessment, total Ti concentrations were determined in muscle, liver, and kidney samples of 29 common dolphins *Delphinus delphis*, five bottlenose dolphins *Tursiops truncatus* and two striped dolphins *Stenella coeruleoalba* stranded on the French Atlantic coast between 1977 and 1990. Ti concentrations did not increase with age, and a temporal comparison did not detect any differences in Ti contamination levels between the 1977–1980 and 1984–1990. Titanium concentrations in almost all samples were below the 0.15 µg g^{-1} dry weight limit of detection, expect for three tissues from three different individuals ranging from 0.22 to 0.83 µg g^{-1} dry weight. The authors concluded that metal concentrations were not significantly decreasing in the North-East Atlantic in the analysed period (Holsbeek et al. 1998). Another study assessed Ti, among other elements, in liver, kidney and muscle tissue of Commerson's dolphins (*Cephalorhynchus c. commersonii*) from subantarctic waters by Instrumental Neutron Activation Analysi, reporting levels below the limit of detection in all analysed samples (Cáceres-Saez et al. 2013). It is important to note, however, that the limits of detection for this analytical technique are relatively high, of 100–150 µg g^{-1} dry weight for liver, 100–250 µg g⁻¹ dry weight for kidney and 40–90 µg g⁻¹ dry weight for muscle in the aforementioned study.

Sperm whale skin levels have also been assessed, reported as ranging from 0.4 to 119.2 µg g⁻¹ dry weight (Wise et al. 2011), while kidney and liver ranged from 0.3 to 1.2 µg g⁻¹ dry weight.

Besides total Ti concentrations, some studies have investigated thermostable subcellular fractions, which comprise a valuable tool for cetacean Ti detoxification assessments (Monteiro et al. 2020). This approach deals with the cellular entry of metals following exposure and is the most adequate way to evaluate toxic effects, as total metal concentrations may still undergo partitioning and/ or detoxification processes (see Hauser-Davis 2023 for a review on the subject). Studies in general are, however, still scarce, even more so for marine mammals. In the first Ti assessment in this regard in marine mammals, five *S. bredanensis* livers, three kidneys and six muscle samples obtained from individuals found stranded in Southeastern Brazil were analysed by inductively coupled plasma mass spectrometry and UV-Vis spectrophotometry, with some MT-Ti associations detected in the thermostable, bioavailable, fraction. (Monteiro et al. 2019), with a limit of quantification of 0.109 mg kg⁻¹ dry weight. However, the authors noted that most subcellular Ti was present in the insoluble and, therefore, non-bioavailable fraction, of all samples, requiring further assessments do evaluate Ti detoxification mechanisms in these animals. In another study by the same research group, total Ti concentrations and metallothionein Ti detoxification were investigated in the Franciscana dolphin *Pontoporia blainvillei* and the Rough-toothed dolphin *Steno bredanensis* from Southeastern Brazil (Monteiro et al. 2020). The authors detected Ti detoxification through MT complexation was observed through size exclusion chromatography coupled to inductively coupled mass spectrometry in thermostable subcellular liver and muscle fractions liver and muscle. Furthermore, TiO₂ NPs contamination was inferred through statistically significant Ti-MT associations, as previous assessments have reported that Ti binds to MT only when in NP form. Furthermore, oceanic TiO₂ diffusion was evidenced, as these MT-Ti correlations were observed for both the coastal (*P. blainvillei*) and offshore (*S. bredanensis*) dolphin species. In

64

addition, Ti detoxification through binding to other detoxification proteins was also verified, such as reduced glutathione, in both species.

Concerning *in vitro* Ti assessments in marine mammals, one study the genotoxic potential of the same TiO_2 NP were assessed *in vitro* Ti in bottlenose dolphin leukocytes from four male and one female specimens reared in captivity in Italy. Three exposure times (4, 24 and 48 h) and three doses (20, 50 and 100 μg mL^{-1}) were investigated. The authors report that both crystalline TiO_2 forms were genotoxic for bottlenose dolphin leukocytes, with significant DNA fragmentation increases following exposure to 50 and 100 μg mL^{-1} for 24 and 48 h (Bernardeschi et al. 2010). In another assessment, the genotoxic potential of TiO_2 anatase NPs and micro-sized rutile on bottlenose dolphins (*Tursiops truncatus*) fibroblasts and leukocytes exposed for 4, 24, and 48 h to 20, 50, 100, 150 μg mL^{-1} TiO_2 was evaluated (Frenzilli et al. 2014). The authors evaluated DNA damage by the single cell gel electrophoresis, also known as the Comet assay. Both types of TiO_2 NPs induced DNA damage, and leukocytes exhibited increased DNA damage after rutile exposure at certain doses/times. An ultrastructural assessment indicated TiO_2 NP cell entry and compartmentalization within membrane-bound vesicles (Frenzilli et al. 2014).

Mammal Models and Humans

Mammalian models, particularly rodents such as mice and rats, are widely used in scientific research and have contributed significantly to the understanding of various biological processes and human health. They are, in fact, crucial in assessing the safety and toxicity of drugs, chemicals, and environmental agents, by aiding in the evaluation of the potential adverse effects of substances on organs, systems, and overall health.

Many mammalian models have been employed, both *in vivo* and *in vitro*, to study the effects of TiO_2 NPs and extrapolate their mechanisms of action and potential toxicity to humans.

One study performed exposures to 20, 50, 100, 150 µg mL^{-1} TiO_2 NPs for 4, 24, and 48 h in human and mouse cell line and investigated DNA damage by single cell gel electrophoresis (Comet assay). Increasing TiO_2 NPs concentrations led to increasing DNA damage in both human and murine fibroblasts, although no linear dose-response effect was observed. Ultrastructural assessments in this case proved that TiO_2 particles entered cells and were compartmentalized within membrane-bound vesicles (Frenzilli et al. 2014).

In another study, the administration of TiO_2 NPs into the abdominal cavity of rats increased the activity of creatine kinase (CK), LDH, AST, and HBDH, as well as increased heart AST, CK and LDH activities, indicating myocardial injury. The authors also reported mitochondrial swelling, increased troponin T, myoglobin, CK-MB and nitric oxide and cardiac tissue caspase-3 (Bu et al. 2010).

In mice, the effects of exposure to 9.38, 18.75, 37.50, 75.00 and 150.00 mg kg^{-1} body weight TiO_2 NPs administered intraperitoneally were evaluated by Bakare et al. (2016). The authors reported increased micronucleus numbers in bone marrow cells of treated mice at 37.50 mg kg^{-1} body weight concentration after 5 days and at all concentrations after 10 days, as well as significantly higher abnormal sperm cells compared to the control group. Furthermore, severe histopathological lesions were also observed, including interstitium oedema congestion, vacuolation and necrosis, suggesting that the bone marrow and testicular cells may be potential targets for TiO$_2$-NPs induced DNA damage and cytotoxicity in mice.

Another assessment investigated the impact TiO$_2$ NPs administered subcutaneously to pregnant female mice, revealing transfer to the offspring and reporting genital and cranial nerve damage to male offspring (Takeda et al. 2009). An energy-dispersive X-ray spectroscopy analysis confirmed the presence of TiO$_2$ nanoparticles in the testis and brain of male mice exposed to these particles at 6 weeks of age. In the subsequent generation of mice born to TiO$_2$-injected parents, the offspring displayed a range of functional and pathological abnormalities, including deleterious alterations in the structure of seminiferous tubules, such as decreased spermatogonia, primary spermatocytes, spermatids, and sperm cells, and increased number of caspase-3 positive cells, an apoptosis biomarker in the olfactory bulb of the brain.

Concerning mechanisms of action, it has been demonstrated that TiO$_2$ NPs directly bind to lactate dehydrogenase (LDH), altering its secondary structure. Thus, it is believed that the nano-anatase TiO$_2$ is responsible for a new metal ion-active site in LDH and subsequently for a LDH activity enhancement (Duan et al. 2009). Furthermore, TiO$_2$ NPs induced apoptosis may be due to increased mitochondrial permeability transition followed by the release of apoptogenic factors such as cytochorome c and activation of caspase-9 and caspase-3 (Faddah et al. 2013). Furthermore, chronic inflammation through B cell activation and IgE production has been noted (Park et al. 2009).

The Microplastics Issue

Microplastics (MPs), defined as particles ranging from 1 to 5,000 µm particles, have become a worldwide concern and an ubiquitous component of our aquatic ecosystems, both marine and freshwater. According to Villarrubia-Gómez et al. (2018), marine plastic contamination is considered irreversible and has a global presence and can now be classified as a planetary boundary threat related to chemical pollution. The amount of microplastics (MPs) in some oceanic compartments is expected, in fact, to double by 2030 and quadrupole by 2060 (see Hale et al. (2020) for a global perspective on microplastics). Furthermore, many plastics are produced by employing metal catalysts or additives, including Ti. This has led to several studies assessing microplastic interactions with Ti and TiO_2-pigmented leaching from microplastics as, in natural environments, MPs undergo various degrees of aging, which can alter their surface properties and leaching behavior compared to unaged MPs (Luo et al. 2020).

For example, one study aimed to investigate the effects of aging on the surface characteristics of TiO_2-pigmented microplastics (MPs) and the particle size distribution of TiO_2 leached from these MPs (Luo et al. 2020). The experimental results demonstrated that aging led to an increase in the carbonyl index and color difference of MPs. This increase in carbonyl content was attributed to photo-oxidation, which resulted in the fragility of MPs, loss of TiO_2 pigments, and the formation of surface cracks. Additionally, the weight loss of aged MPs was reduced due to the photo-transformation of low-density polyethylene (LDPE) into other compounds. Furthermore, aged MPs exhibited a higher and faster release of TiO_2 particles compared to unaged MPs. Aged MPs tended to release more large particles (> 5 μm) and fewer small particles (< 5 μm) compared to unaged MPs. The particle size distribution of TiO_2 leached from aged MPs was uniformly distributed across each size fraction, while unaged MPs predominantly released small particles (57.6%–86.2%). As the leaching time increased, the proportion of small particles leached from MPs, particularly aged MPs, decreased, while the proportion of large particles increased. These findings enhance our understanding of how the aging process affects the properties and leaching behavior of MPs, as well as the size distribution of particulate additives leached from MPs.

However, against this trend, some studies have also assessed microplastic against toxic Ti effects. For example, Thiagarajan et al. (2019a) evaluated the influence of differently functionalized polystyrene microplastics on the toxic effects of P25 TiO_2 NPs towards marine algae *Chlorella* sp. Aggregation of TiO_2 NPs in artificial seawater was observed to increase over time with a higher toxic effect on algae. On the other hand, regardless of their functionalization, microplastics had minimal toxic effects, only causing a 15% effect at the highest tested concentration of $1000\,mg\,L^{-1}$. However, the toxicity of TiO_2 NPs was enhanced in the presence of plain and aminated polystyrene microplastics, as confirmed by oxidative stress determination studies such as reactive oxygen species and lipid peroxidation assays. In contrast, carboxylated polystyrene microplastics, which were negatively charged, decreased the toxicity of TiO_2 NPs, which the authors postulate as probably due to heteroaggregation between the two materials in the system.

Another assessment in this regard aimed to investigate the impact of different surface-functionalized polystyrene microplastics (carboxylated, plain, and aminated) on the toxicity of TiO_2 NPs, with an additional assessment conducted involving two trophic levels: *Chlorella* sp. as the prey and *Artemia salina* as the predator., providing insights into the toxicity at different trophic levels and the transfer potential of TiO_2 NPs in the presence of various microplastics (Thiagarajan et al. 2019b). Several experiments were conducted on *Chlorella* sp., including toxicity assessment, oxidative stress determination, and evaluation of TiO_2 NPs uptake in the presence and absence of microplastics. The findings indicated that aminated and plain polystyrene microplastics intensified TiO_2 NPs toxicity in *Chlorella* sp., whereas carboxylated microplastics reduced its toxic effects. Concerning *Artemia salina*, the toxicity assessment was performed using two exposure methods: aqueous and dietary routes. The aqueous route, which involved direct exposure to TiO_2 NPs and microplastics, demonstrated greater toxicity, uptake, and accumulation in *Artemia salina* compared to the dietary route. Notably, dietary exposure reduced the toxicity, uptake, and accumulation of nano-TiO^2. No significant change in the biomagnification factors of TiO_2 NPs was observed for all tested concentrations, whether combined with or without microplastics. The computed values were consistently less than 1, indicating minimal transfer of TiO_2 NPs from *Chlorella* sp. to *Artemia salina*.

Ecological Effects
Titanium Bioaccumulation and Biomagnification Processes throughout Aquatic Food Webs

Many metals, including Ti, have been noted as undergoing both bioaccumulation and biomagnification processes. Bioaccumulation refers to the gradual accumulation and concentration of substances, such as pollutants or chemicals, in living organisms over time, when organisms take in these compounds at rates faster their elimination. As a result, the concentration of the substance increases in the organism's tissues or organs. Biomagnification, on the other hand, is the process by which the concentration of certain substances increases at higher levels of a food chain or food web. Unlike bioaccumulation, which refers to the accumulation of substances within an individual organism, biomagnification refers to the increase in concentration as one moves up the trophic levels in an ecosystem. Interestingly, metals can exhibit threshold effects, especially in aquatic ecosystems, where the ecological impacts become more pronounced above a certain concentration (Rodrigues et al. 2022). Below the threshold, organisms may be able to tolerate and recover from metal exposure. However, once the threshold is exceeded, the effects can become more severe and long-lasting.

The ecological effects of Ti, as with other metals, can vary depending on its form, concentration, and exposure pathways. Although several studies have indicated Ti uptake, mainly in NP form, and transfer along aquatic food chains in laboratory experiments, these trials often consist of only two or three model organisms (Kalman et al. 2015, Wang et al. 2017), and limited data is available concerning to Ti bioaccumulation and biomagnification processes in real-life scenarios throughout complex food webs (Xiao et al. 2019).

Concerning laboratory assessments, a simplified food chain experiment reported TiO_2 NPs transfer from *Daphnia* to zebrafish in a simplified food chain (Zhu et al. 2010), although with biomagnification factors below one, indicating no TiO_2 NPs biomagnification. However, even with no detected biomagnification, zebrafish were also able to accumulate TiO_2 NPs by aqueous exposure, with extremely high bioconcentration factors of 25.38 and 181.38 following exposure to 0.1 mg L^{-1} and 1.0 mg L^{-1} TiO_2 NPs, respectively. These authors, thus indicate that dietary intake may comprise a major nanomaterial exposure route for higher aquatic trophic levels, although aqueous exposure is also important. Similarly, another investigation noting increasing TiO_2 NPs accumulation and trophic transfer from clamworms *Perinereis aibuhitensis* to juvenile turbots (flatfish) *Scophthalmus maximus* following exposure to 10, 50 and 100 mg L^{-1} TiO_2 NPs (Wang et al. 2016), also indicating negative growth effects and abnormal liver and spleen symptoms. That study also reported higher Ti accumulation in the highest trophic level, in this case, turbots, through waterborne exposure compared to the dietary route. In another study, trophic TiO_2 NPs transfer from the marine microalga *Nitzschia closterium* to the Farrer's scallop *Chlamys farreri* was verified, with similar waterborne x trophic transfer findings, higher in the former compared to the latter. Deleterious scallop effects were also reported, such as lysosomal membrane permeability, DNA damage, and histopathological effects. In another experimental assessment, the toxic and cumulative effects of TiO_2 NPs on an aquatic food chain were investigated, focusing on two trophic levels: producers (algae) and consumers (rotifers) (Li et al. 2022). The results demonstrate that exposure to TiO_2 NPs suspensions had detrimental effects on both producers and consumers. A concentration-dependent reduction in the density of algal cells caused by nanoparticles. Furthermore, the hatching life expectancy, average lifespan, net reproductive rate, and population intrinsic growth rate of rotifers significantly decreased with increasing concentrations of nanomaterials ($P < 0.05$). Importantly, TiO_2 NPs were found to accumulate in algal cells and were subsequently transferred to consumers through dietary exposure. This resulted in biomagnification of TiO_2 NPs within the simplified food chain, as indicated by biomagnification factor (BMF) values exceeding 1 in many cases. The accumulation of nanoparticles in algae and rotifers was influenced by exposure concentration, time, and interactions.

On the other hand, some studies have indicated Ti biodilution (both in total and in NP form, ranging from 46.6 to 116 nm) throughout food webs, for example, in a natural aquatic food web in Taihu Lake, China (Xiao et al. 2019). The findings of that study indicated trophic magnification factors below one, although both Ti forms were detected in the water, sediment, and aquatic organisms in the study area. The authors indicate that the sediment was the main sink for NP Ti and one of the most important exposure sources for local invertebrates.

Therefore, ecosystem-level studies are paramount to further investigate how Ti and TiO_2 NPs transfer throughout trophic levels, leading to both ecological and human health effects.

Interactive Titanium Effects with Other Contaminants

The continuous and exorbitant inputs of Ti, mainly in the form of TiO_2 NP, into aquatic systems has led to concern not only on the toxicity of these compounds alone, but also regarding their potential interactions with other environmental pollutants, both organic and inorganic (Canesi et al. 2014), in different environmental compartments (water column, sediment, biota), and potentially throughout aquatic trophic webs, comprising an increasing concern for exposed biota (Rocco et al. 2015). This is due to certain TiO_2 NP properties, such as high specific surface area and number of surface activation sites, and therefore, as well as high sorption capacity (Chen et al. 2011). The specific outcomes of these interactions depend on the nature of the contaminants and the environmental conditions.

Interactions between variables can be categorized as additive, synergistic, or antagonistic. These are defined by Fong et al. (2018) as "additive when their combined effect is the sum of each independently, synergistic when the combined effect is greater than the sum of each independently, and antagonistic when the combined effect is less than the sum of each independently" (Fong et al. 2018). It is important to note that the type of interaction (additive, synergistic, or antagonistic) depends on various factors, such as the specific contaminants involved in the process, their concentrations, exposure duration, environmental conditions, and the sensitivity of the organisms or ecosystems being affected (Fischer et al. 2013). Additionally, interactions between contaminants may vary across different biological endpoints and environmental contexts.

Therefore, assessing the combined effects of contaminants, including titanium, requires careful consideration of these factors and the specific circumstances of the exposure scenario.

In this regard, some studies have demonstrated increased toxicity following concomitant exposure to TiO_2 NPs and other metals. For example, one study assessed effects in the green algae *Scenedesmus obliquus* exposed to TiO_2 NP and Cd^{2+} at different concentration ratios (Wang et al. 2021). Three different combined toxicity modes were observed, namely antagonistic, partially additive, and synergistic. The antagonistic toxicity effect was observed under co-exposure to low concentration ratio of the evaluated contaminants due to Cd^2 absorption by TiO_2 NP, decreasing Cd^2 bioavailability while the partially additive and synergistic toxicities were observed at relatively high co-exposure concentrations, mechanically and/or oxidatively damaging algae cell structures, where TiO_2 NP enhanced the amount of Cd^{2+} entering cells, resulting in higher toxicity. In view of these results, the authors indicate that the TiO_2 NP and Cd^{2+} concentration ratios play an important role in determining the combined toxicity mode of these contaminants. In another assessment, the effects of 2 mg L^{-1} TiO_2 NP on tributyltin (TBT) toxicity, a highly toxic antifouling compound, in abalone *Haliotis diversicolor supertexta* embryos was evaluated (Zhu et al. 2011). No developmental effects were observed at 2 mg L^{-1} TiO_2 NP, while concentrations over 10 mg L^{-1} TiO_2 NP resulted in hatching inhibition and malformations. Co-exposure of 2 mg L^{-1} TiO_2 NP and TBT increased TBT up to 20-fold compared to TBT alone in an additive interaction, which the authors postulate as being due to the combined effects of TBT adsorption onto TiO_2 NP and TiO_2 NP internalization by abalone embryos. In another study, TiO_2 NPs were noted as Cd carriers at sub-lethal concentrations in different environmental compartments to the protozoan *Tetrahymena thermophila* (Yang et al. 2014), the water flea, *Daphnia magna*, which inhabits the water column, and the California blackworm *Lumbriculus variegatus*, which inhabits the sediment (Hartmann et al. 2012). Increased Ag and Cu toxicity to Daphnia magna following exposure to TiO_2 NPs have also been reported (Rosenfeldt et al. 2014, Fan et al. 2011), as well as enhanced Cd and arsenate bioaccumulation in carp *Cyprinus carpio* tissues through interactions with TiO_2 NPs (Sun et al. 2007, Zhang et al. 2007).

Another assessment investigated the effects of TiO_2 NPs on the toxicity of phenanthrene (Phe) and Cadmium (Cd^{2+}) to *Artemia salina*, a marine zooplankton species used as a model organism (Lu et al. 2018). The TiO_2 NPs exhibited limited toxicity to *A. salina* when tested alone within a 48-hour exposure period. However, a significant impact of TiO_2 NPs on the toxicity of Phe and Cd^{2+} to A. salina was noted. In the presence of 5 mg L^{-1} TiO_2 NPs, the toxicity of both compounds increased by 2.0% and 12.2%, respectively, compared to their individual toxicities. However, when TiO_2 NPs concentrations were increased to 400 mg L^{-1}, the toxicity of Phe and Cd^{2+} decreased by 24.5% and 57.1%, respectively, in a dose-dependent manner. The authors indicate that several factors contributed to the concentration-dependent impacts of TiO_2 NPs on the toxicity of both evaluated pollutants, namely pollutant adsorption on the TiO_2 NPs, the facilitated bioaccumulation in the presence of TiO_2 NPs, the limited gut volume of organisms, and the aggregation and sedimentation behaviors of TiO_2 NPs.

Another study assessed the uptake, assimilation efficiency (AE), and efflux rate constant (ke) of two metals, Cd and Zn, adsorbed on TiO_2 NPs were in the freshwater zooplankton *Daphnia magna* (Tan et al. 2012). The biokinetics of these metals in daphnids exposed to TiO_2 NPs-bound metals were compared to those exposed only to dissolved metals as controls. Water Cd and Zn uptake involved an initial rapid uptake followed by apparent saturation. The observed metal uptake was accompanied by ingestion of TiO_2 NPs. The AE of Cd and Zn adsorbed on the TiO_2 NPs was 24.6 ± 2.4% and 44.5 ± 3.7%, respectively, while for dissolved metals the AE was 30.4 ± 3.4% for Cd and 51.8 ± 5.0% for Zn. The AE of adsorbed metals decreased with increasing concentrations of TiO_2 NPs. The authors report that the observed difference in AE between Cd and Zn suggests that desorption of these metals from TiO_2 NPs may occur within daphnid gut. When TiO_2 NPs were carried by algae, the AE of Cd and Zn adsorbed on to the TiO_2 NPs was significantly higher compared to direct exposure to nano-TiO_2. Additionally, the efflux rate constants of Cd and Zn adsorbed on TiO_2 NPs in the zooplankton were significantly lower than those of Cd and Zn not adsorbed on TiO_2 NPs. The findings demonstrate that Cd and Zn uptake and retention are enhanced when these metals are adsorbed on nano-TiO_2 particles, highlighting the potential influence of TiO_2 NPs on the bioavailability and toxicity of other contaminants.

Another study aimed to investigate the potential modulation of toxic effects following interaction of TiO_2 NPs with Cr(VI) in the presence of the marine crustacean, *Artemia salina* (Thiagarajan et al. 2020). The TiO_2 NPs formed agglomerates in artificial sea water (ASW), and settled as micron-sized particles. The addition of Cr(VI) to the mixture resulted in increased agglomeration due to Cr(VI) sorption onto nano-TiO_2, leading to a decrease in the remaining Cr concentrations in the suspension. Acute toxicity tests were conducted using pristine TiO_2 NPs (0.25, 0.5, 1, 2, and 4 mg L^{-1}) and Cr(VI) (0.125, 0.25, 0.5, and 1 mg L^{-1}), demonstrating a concentration-dependent increase in *Artemia salina* mortality. To examine the effects of mixtures of TiO_2 NPs and Cr(VI) on *Artemia salina*, two sets of experiments were conducted. The first investigated the toxic effect of TiO_2 NPs (0.5, 1, 2, and 4 mg L^{-1}) in combination with a fixed concentration (0.125 mg L^{-1}) of Cr(VI). The second group studied the toxicity of Cr(VI) (0.25, 0.5, and 1 mg L^{-1}) with a fixed concentration (0.25 mg L^{-1}) of TiO_2 NPs. The results indicated that the toxic effects of TiO_2 NPs were not significantly reduced at a fixed concentration of Cr(VI). However, there was a significant reduction in the toxicity of Cr(VI) when a fixed concentration of TiO_2 NPs was present. These findings were supported by an independent action model, which demonstrated an antagonistic mode of action between nano-TiO_2 and Cr(VI). Additionally, the generation of reactive oxygen species (ROS) and the measurement of antioxidant enzyme activities aligned with the toxicity results. The authors indicate that their study suggests that modifying the toxicity of Cr(VI) at a fixed concentration of nano-TiO_2 could have a significant impact on reducing Cr(VI) toxicity across different trophic levels.

Interactions between several contaminants and TiO_2 NPs concerning trophic transfer processes have also been investigated, with conflicting results. One investigation in this regard assessed the bioaccumulation profile and feeding behaviour effects of TiO_2 NPs and Pb, on copepods within a simplified food chain consisting of the freshwater alga *Chlorella ellipsoides* and cyclopoid copepods (Matouke and Mustapha 2018). The findings indicate that both Pb and TiO_2 NPs, either individually or in combination, were transferred from the alga to the copepods through the dietary pathway. The highest bioconcentration factor (BCF) of 748.5 was observed for Pb in the combined compound treatment (Pb 15 + Ti 16.5 µg L^{-1}), while the highest BCF of 5.57 was recorded for TiO_2 NPs in the TiO_2 NPs (16.5 µg L^{-1}) treatment alone. Ingestion and filtration rates were significantly decreased in all treatment groups, with a significant reduction in these rates in the presence of the combined metals. Furthermore, the individual compounds as well as their combinations significantly increased carbohydrate and total lipid content in copepods. Regarding antioxidant activities, significant effects were observed for Glutathione peroxidase (GPx), Glutathione reductase (GR), and catalase (CAT), while superoxide dismutase (SOD) and malondialdehyde (MDA) levels were not significantly affected by both single and binary treatments. Overall, the results demonstrate that the co-exposure of TiO_2 NPs and Pb inhibits the ingestion and filtration of microalgae by cyclopoid copepods and induces an increase in carbohydrate, lipid, and antioxidant enzyme activities such as GPx, GR, and CAT, indicating a stress response.

Trophic transfer processes have, however, been noted in other studies. For example, one study in this regard assessed the behavior of Cr, Cu, Pb and Se adsorbed to TiO_2 NPs throughout a three trophic level aquatic food chain comprising *Ceratium tripos* (Müller) (Nitzsch 1817) as the phytoplankton, *Daphnia hyalina* (Leydig 1860) as the zooplankton and *Liza abu* (Heckel 1843) as the predator (Hosseini et al. 2015). The phytoplankton were exposed to 0, 0.2 and 0.5 µg L^{-1} TiO_2 NPs and fed to the zooplankton which was then used as food for the predator. Significant differences between Cr, Pb and Se among the trophic groups were noted, with Cr and Se biomagnifying through the food chain.

Some of the reported synergistic interactions may be due to a significant increase in TCDD accumulation in whole mussels in the presence of TiO_2 NP, through the "Trojan horse effect". The Trojan horse effect refers to a phenomenon in which n some contaminants act as carriers or vehicles for the uptake of others potentiating toxic effects (Zhang and Goss 2021). The term derived from the ancient Greek myth where a giant wooden horse was used to conceal Greek soldiers and infiltrate the city of Troy.

Some studies, however, have reported conflicting results concerning coexposure to TiO_2 NPs. For example, Nigro et al. (2015) investigated various genotoxic endpoints in the European sea bass (*Dicentrarchus labrax*) exposed to TiO_2 NPs (1 mg L^{-1}) alone and in combination with $CdCl_2$ (0.1 mg L^{-1}) for 7 days. The authors assessed primary DNA damage using the comet assay, identified apoptotic cells through the diffusion assay, and examined the occurrence of micronuclei and nuclear abnormalities using the cytome assay in peripheral erythrocytes. They also evaluated genomic stability in muscle tissue using the random amplified polymorphism DNA-PCR (RAPD) assay. Concerning effects, TiO_2 NPs alone induced chromosomal alterations but had limited effects on DNA damage, while $CdCl_2$ exposure primarily caused DNA damage rather than chromosomal alterations. Coexposure to TiO_2 NPs and $CdCl_2$, however, appeared to mitigate chromosomal damage and partially restore genome template stability.

Antagonistic interactions have also been reported, where TiO_2 NPs significantly decreased inhibitory Cd effects towards *Chlamydomonas reinhardtii* (Yu et al. 2018), Cu and As toxicity towards *Daphnia magna* (Rosenfeldt et al. 2014). This mechanism has been postulated as due to co-contaminant adsorption to TiO_2 NPs, decreasing their concentrations in solution and alleviating their toxic effects (Dutta et al. 2004, Yu et al. 2018). This has been postulated as due to high TiO_2 NP concentrations, as TiO_2 NPs suspensions in these conditions become is unstable and may precipitate following aggregation which may decrease free metal ion levels in solution, and lower their toxicity, although this mechanism has not yet been fully elucidated (Wang et al. 2021).

Concerning organic contaminants, one study assessed the potential interactive effects of TiO_2 NP and the organic compound 2,3,7,8-Tetrachlorodibenzo-p-dioxin (2,3,7,8-TCDD), a polychlorinated dibenzo-p-dioxin, in *Mytilus galloprovincialis* mussels (Canesi et al. 2014). *In vitro* experiments were carried out employing TiO_2 NP and TCDD both alone and in combination at different contaminant concentrations and exposure times. Mussels were also exposed *in vivo* to *both contaminants at* 100 µg L^{-1} (TiO_2 NP) and 0.25 µg L^{-1} (2,3,7,8-TCDD), alone and combined for 96 h. Significant effects on a wide range of biomarkers, from the molecular to the tissue level, were observed in both cases. The *in vitro* experiments indicated antagonistic effects of on phagocytic activity of mussel haemocytes and on efflux activities mediated by ABC transporters in gills, while the *in vivo* assays indicated antagonistic effects on haemocyte phagocytic activity and ABC mediated efflux activities in gills and synergistic effects at the tissue level on lysosome/cytoplasm volume ratios in the digestive gland, as well as on transcription of estrogen receptors.

Concerning the same organic compound, another study evaluated the influence of TiO_2 NP on 2,3,7,8-TCDD bioconcentration and toxicity in the marine fish European sea bass (*Dicentrarchus labrax*) during a 7-day *in vivo* exposure (Della Torre et al. 2015). A multimarker approach was employed to assess various aspects in different organs, namely liver for detoxification, gills and spleen for innate immunity, pro-inflammatory response, and adaptive immunity, peripheral erythrocytes for genotoxicity, and muscle for bioconcentration. The bioconcentration of 2,3,7,8-TCDD in the presence of TiO_2 NP in the liver, skin, and muscle, as well as the interaction between TiO_2 NP and organic pollutants in artificial sea water (ASW). The reported findings revealed that TiO_2 NP had a negative influence on the immune response induced by 2,3,7,8-TCDD in the spleen but not in the gills. Additionally, TiO_2 NP reduced the DNA damage caused by 2,3,7,8-TCDD in erythrocytes. However, no interference of TiO_2 NP with 2,3,7,8-TCDD detoxification and bioconcentration was noted, as there no interaction took place between TiO_2 NP and the investigated organic pollutant in ASW.

In another study, the joint toxicities of TiO_2 NP alongside co-exposure to four different organochlorine contaminants

(OC) (atrazine, hexachlorobenzene, pentachlorobenzene, and 3,3′,4,4′-tetrachlorobiphenyl) in *Chlorella pyrenoidosa* algae (Zhang et al. 2017). Contaminant coexposure led to a synergistic effect between TiO_2 NP andatrazine, an antagonistic effect between TiO_2 NP and both hexachlorobenzene and 3,3′,4,4′-tetrachlorobiphenyl, and an additive effect between TiO_2 NP and pentachlorobenzene. Almost no OC adsorption by TiO_2 NP was observed, and the physicochemical properties of the NP were mostly unaltered by the presence of OC. However, both NP and OC significantly affected the biophysicochemical properties of algal cells, influencing cell surface binding and/or internalization. TiO_2 NP significantly increased the bioaccumulation of each OC individually, although TiO_2 NP bioaccumulation following co-exposure to all OC except atrazine.

Concerning interactions with pharmaceuticals, one assessment evaluated the interactive effects between TiO_2 NP and tetracycline (TC), an antibiotic in a freshwater alga, *Scenedesmus obliquus* (Iswarya et al. 2017). Based on the calculated TiO_2 NP EC_{50} of 136.88 ± 2.30 μmol L^{-1}, three different TC concentrations (0.34, 0.68, and 1.36 μmol L^{-1}) were selected to investigate toxicity in algae exposed to 18.75, 37.5, and 75 μmol L^{-1} TiO_2 NP. Algae growth inhibition was noted at lower TiO_2 NPs concentrations alongside TC, while decreased growth inhibition was noted with increasing contaminant concentrations. Both additive and antagonistic effects were observed. Furthermore, TC exposure lowered TiO_2 NP uptake by algal cells, indicating TC dominance concerning toxic additive or antagonistic TiO_2 NPs.

Climate Change Considerations

Some studies have been cited concerning climate change scenario simulations for example, employing mollusks exposed to TiO_2 NP. Studies on TiO_2 NP kinetics under simulated climate change conditions, however, are still lacking, and no cohesive assessments are available in this regard. It is, however, important to note that one of the most noticeable climate change effects comprises increases in global temperatures. This effect will directly affect aquatic environments, as temperature changes have the potential to modulate the toxicity of emerging contaminants, including NPs (Zhang et al. 2022). In a recent review, Zhang et al. (2022) compiled and examined the effects of temperature on the toxicity of waterborne NPs across various organisms, reporting a consistent trend: temperature increases result in a general reduction in the bioavailability of NPs in algae and plants, primarily due to increased aggregation. As a consequence, the overall toxicity of NPs tends to decrease with temperature elevation due to agglomeration of larger particles. However, agglomeration can create shading effects, impeding algal photosynthesis. Conversely, the toxicity of NPs in microorganisms and aquatic animals tends to increase with rising temperatures, which the authors primarily attribute to the significant influence of high temperatures on chemical uptake and excretion, leading to increased production of reactive oxygen species and heightened oxidative damage within organisms. Furthermore, elevated temperatures can affect the formation and composition of a protein corona on NPs, thereby altering their toxicity profiles.

Conclusions

The concentrations of Ti, a contaminant of emerging concern, in living organisms are still scarce for many groups, mostly conducted within laboratory settings, with a lack of evaluations conducted within the environmental monitoring assessment framework. This is especially true concerning trophic transfer and potential biomagnification processes, as evaluations in this regard are almost non-existent in real-life scenarios. Of note, a significant knowledge gap is noted concerning this element's subcellular compartmentalization and its sublethal and/or lethal ecotoxicological effects. Further assessments should focus on assessing potentially deleterious intracellular and physiological effects to understand this elements toxicokinetics and toxicodynamics in exposed organisms. The reported Ti concentrations in different invertebrate and vertebrate groups to date, although still limited, which comprise valuable data for establishing baseline levels in pristine environments and contamination values in polluted areas, which may be, in turn, applied to conservation issues. In a Public Health setting, although Ti was for many years considered safe and permitted as a food additive worldwide, the scenario has changed significantly in the last decades, as more evidence has been amassed concerning this element's potential toxic effects, including genotoxicity. This further underlines the need for additional ecotoxicological and toxicological assessments concerning this element.

Acknowledgements

This research was funded by the Carlos Chagas Filho Foundation for Research Support of the State of Rio de Janeiro (FAPERJ) (RAHD), through a Jovem Cientista do Nosso Estado 2021–2024 grant (process number E-26/201.270/2021), Jovem Pesquisador Fluminense com Vínculo em ICTS do Estado do Rio de Janeiro grant (process number E-26/2010.300/2022), and the Brazilian National Council of Scientific and Technological Development (CNPq), through a productivity grant (process number 308811/2021-6). The implementation of the Projeto Pesquisa Marinha e Pesqueira is a compensatory measure established by the Conduct Adjustment Agreement under the responsibility of the PRIO company, conducted by the Federal Public Ministry-MPF/RJ.

References

Adam, V., S. Loyaux-Lawniczak, J. Labille, C. Galindo, M. del Nero, S. Gangloff, T. Weber and G. Quaranta. 2016. Aggregation behaviour of TiO_2 nanoparticles in natural river water. J. Nanopart. Res. 18: 13.

Adamu, Y.A., A. Usman, U.M. Mera, M.B. Abubakar, A. Bello and M.A. Umaru. 2015. Haematological values of broilers managed on titanium dioxide treated litter. International Invention Journal of Medicine and Medical Sciences. 2: 139–143.

Ahmed, N.M. and A.M. Taha. 2022. Histopathological effects of titanium dioxide nanoparticles on the liver of Japanese quail *Coturnix coturnix japonica*. Iraqi Journal of Veterinary Sciences. 36: 349–358.

Ahmed, N., K. Tanveer, Z. Younas, T. Yousaf, M. Ikram, N.I. Raja, Z.-R. Mashwani, S. Alghamdi, I.S. Al-Moraya and N.T. Shesha. 2023. Green-processed nano-biocomposite (ZnO–TiO_2): Potential candidates for biomedical applications. Green Process. Synth. 12: 1.

Al Mahrouqi, D., S. Al Riyami and M.J. Barry. 2018. Effects of Zn and Ti nanoparticles on the survival and growth of *Sclerophrys arabica* tadpoles in a two level trophic system. B. Environ. Contam. Tox. 101: 592–597.

Alexander, L., A. Agyekumhene and P. Allman. 2017. The role of taboos in the protection and recovery of sea turtles. Front. Mar. Sci. 4: 237.

Al-Jomily, Z. and R.G. Al-Sultan. 2022. Effect of nano scale titanium dioxide in quail testes tissue and molecular diagnosis in a RAPD_PCR method. HIV Nursing. 22: 2236–2239.

Alla, R.K., K. Ginjupalli, N. Upadhya, M. Shammas, R. Krishna and R. Sekhar. 2011. Surface roughness of implants: A review. Trends Biomater. Artif. Organs. 25: 112–118.

Aroche, R., Y. Martínez, Z. Ruan, G. Guan, S. Waititu, C.M. Nyachoti, D. Más and S. Lan. 2018. Dietary inclusion of a mixed powder of medicinal plant leaves enhances the feed efficiency and immune function in broiler chickens. J. Chem. e4073068.

Aronsson, B.O., J. Lausmaa and B. Kasemo. 1997. Glow discharge plasma treatment for surface cleaning and modification of metallic biomaterials. J. Biomed. Mater. Res. 35: 49–73.

Bai, R.G. 2015. Technology and development of high-efficiency clean utilization of vanadium–titanium ore resource in Cheng steel. Hebei Metall. 12: 1.

Bakare, A.A., A.J. Udoakang, A.T. Anifowoshe, O.M. Fadoju, O.I. Ogunsuyi, O.A. Alabi, C.G. Alimba and I.T. Oyeyemi. 2016. Genotoxicity of titanium dioxide nanoparticles using the mouse bone marrow micronucleus and sperm morphology assays. Journal of Pollution Effects & Control. 4: 1–7.

Batley, G.E. and P.G.C. Campbell. 2022. Metal contaminants of emerging concern in aquatic systems. Environ. Chem. 19: 23–40.

Behrenfeld, M.J., R.T. O'Malley, D.A. Siegel, C.R. McClain, J.L. Sarmiento, G.C. Feldman, A.J. Milligan, P.G. Falkowski, R.M. Letelier and E.S. Boss. 2006. Climate-driven trends in contemporary ocean productivity. Nature. 444: 752–755.

Bernardeschi, M., P. Guidi, V. Scarcelli, G. Frenzilli and M. Nigro. 2010. Genotoxic potential of TiO_2 on bottlenose dolphin leukocytes. Anal. Bioanal. Chem. 396: 619–623.

Bocchetta, P., L.-Y. Chen, J.D.C. Tardelli, A.C. Reis, F. Almeraya-Calderón and P. Leo. 2021. Passive layers and corrosion resistance of biomedical Ti-6Al-4V and β-Ti alloys. Coatings. 11: 487.

Boobis, A.R., S.M. Cohen, V.L. Dellarco, J.E. Doe, P.A. Fenner-Crisp, A. Moretto, T.P. Pastoor, R.S. Schoeny, J.G. Seed and D.C. Wolf. 2016. Classification schemes for carcinogenicity based on hazard-identification have become outmoded and serve neither science nor society. Regul. Toxicol. Pharm. 82: 158–166.

Bosio, M., S. Satyro, J.P. Bassin, E. Saggioro and M. Dezotti. 2019. Removal of pharmaceutically active compounds from synthetic and real aqueous mixtures and simultaneous disinfection by supported TiO_2/UV-A, H_2O_2/UV-A, and TiO_2/H_2O_2/UV-A processes. Environ. Sci. Pollut. R. 26: 4288–4299.

Bostock, J., A. Lane, C. Hough and K. Yamamoto. 2016. An assessment of the economic contribution of EU aquaculture production and the influence of policies for its sustainable development. Aquacult. Int. 24: 699–733.

Botta, C., J. Labille, M. Auffan, D. Borschneck, H. Miche, M. Cabié, A. Masion, J. Rose and J. Bottero. 2011. TiO_2-based nanoparticles released in water from commercialized sunscreens in a life-cycle perspective: Structures and quantities. Environ. Pollut. 159: 1543–1550.

Boutillier, S., S. Fourmentin and B. Laperche. 2022. History of titanium dioxide regulation as a food additive: A review. Environ. Chem. Lett. 20: 1017–1033.

Bouwman, H., P. Booyens, D. Govender, D. Pienaar and A. Polder. 2014. Chlorinated, brominated, and fluorinated organic pollutants in Nile crocodile eggs from the Kruger National Park, South Africa. Ecotox. Environ. Safe. 104: 393–402.

Boyer, R.R. 1996. An overview on the use of titanium in the aerospace industry. Mat. Sci. Eng. A-Struct. 213: 103–114.

Boyer, R.R. 2010. Attributes, characteristics, and applications of titanium and its alloys. JOM. 62: 21–24.

Bracamontes-Ruelas, A.R., L.A. Ordaz-Díaz, A.M. Bailón-Salas, J.C. Ríos-Saucedo, Y. Reyes-Vidal and L. Reynoso-Cuevas. 2022. Emerging Pollutants in Wastewater, Advanced Oxidation Processes as an Alternative Treatment and Perspectives. Processes 10: 1041.

Branen, A.L., P.M. Davidson, S. Salminen and J. Thorngate. 2001. Food Additives. CRC Press, Boca Raton.

Britannica, The Editors of Encyclopaedia. 2022. Titanium. Encyclopedia Britannica, 2 Sep. 2022. Available at: https://www.britannica.com/science/titanium. (Accessed 25 December 2022).

Brunet, J.-C. and C. Hauviller. 1977. A titanium high pressure gas target. Nucl. Instrum. Methods. 153: 59–60.

Bu, Q., G. Yan, P. Deng, F. Peng, H. Lin, Y. Xu, Z. Cao, T. Zhou, A. Xue and Y. Wang. 2010. NMR-based metabonomic study of the sub-acute toxicity of titanium dioxide nanoparticles in rats after oral administration. Nanotechnology. 21: 125105.

Cáceres-Saez, I., S. Ribeiro Guevara, N.A. Dellabianca, R.N.P. Goodall and H.L. Cappozzo. 2013. Heavy metals and essential elements in Commerson's dolphins (*Cephalorhynchus c. commersonii*) from the southwestern South Atlantic Ocean. Environ. Monit. Assess. 185: 5375–5386.

Canesi, L., G. Frenzilli, T. Balbi, M. Bernardeschi, C. Ciacci, S. Corsolini, C. Della Torre, C., R. Fabbri, C. Faleri, S. Focardi, P. Guidi, A. Kočan, A. Marcomini, M. Mariottini, M. Nigro, K. Pozo-Gallardo, L. Rocco, V. Scarcelli, A. Smerilli and I. Corsi. 2014. Interactive effects of n-TiO$_2$ and 2,3,7,8-TCDD on the marine bivalve Mytilus galloprovincialis. Aquat. Toxicol. 153: 53–65.

Capodaglio, A.G. 2020. Critical perspective on advanced treatment processes for water and wastewater: AOPs, ARPs, and AORPs. Appl. Sci. 10: 4549.

Carmo, T.L.L., P.R. Siqueira, V.C. Azevedo, D. Tavares, E.C. Pesenti, M.M. Cestari, C.B.R. Martinez and M.N. Fernandes. 2019. Overview of the toxic effects of titanium dioxide nanoparticles in blood, liver, muscles, and brain of a Neotropical detritivorous fish. Environ. Toxicol. 34: 457–468.

Chen, J., X. Dong, Y. Xin and M. Zhao. 2011. Effects of titanium dioxide nano-particles on growth and some histological parameters of zebrafish (*Danio rerio*) after a long-term exposure. Aquat. Toxicol. 101: 493–499.

Chen, T., J. Yan and Y. Li. 2014. Genotoxicity of titanium dioxide nanoparticles. Journal of Food and Drug Analysis. 22: 95–104.

Chen, Y., X. Tang, X. Gao, B. Zhang, Y. Luo and X. Yao. 2019. Antimicrobial property and photocatalytic antibacterial mechanism of the TiO$_2$-doped SiO$_2$ hybrid materials under ultraviolet-light irradiation and visible-light irradiation. Ceram. Int. 45: 15505–15513.

Chen, D., Y. Cheng, N. Zhou, P. Chen, Y. Wang, K. Li, S. Huo, P. Cheng, P. Peng, R. Zhang, L. Wang, H. Liu, Y. Liu and R. Ruan. 2020a. Photocatalytic degradation of organic pollutants using TiO$_2$-based photocatalysts: A review. J. Clean. Prod. 268: 121725.

Chen, Z., S. Han, S. Zhou, H. Feng, Y. Liu and G. Jia. 2020b. Review of health safety aspects of titanium dioxide nanoparticles in food application. NanoImpact. 18: 100224.

Cheng, X., H. Liu, Q. Chen, J. Li and P. Wang. 2014. Preparation of graphene film decorated TiO_2 nano-tube array photoelectrode and its enhanced visible light photocatalytic mechanism. Carbon. 66: 450–458.

Church, N.L., E.M. Hildyard and H.G. Jones. 2021. The influence of grain size on the onset of the superelastic transformation in Ti–24Nb–4Sn–8Zr (wt%). Mat. Sci. Eng. A-Struct. 828: 142072.

Clark, J. and A. Williams-Jones. 2004. Rutile as a potential indicator mineral for metamorphosed metallic ore deposits. Rapport Final de DIVEX, Sous-projet SC2, Montréal, Canada.

Clemente, Z., V.L. Castro, L.O. Feitosa, R. Lima, C.M. Jonsson, A.H.N. Maia and L.F. Fraceto. 2013. Fish exposure to nano-TiO_2 under different experimental conditions: Methodological aspects for nanoecotoxicology investigations. Sci. Total Environ. 463–464: 647–656.

Cobelo-García, A., M. Filella, P. Croot, C. Frazzoli, G. Du Laing, N. Ospina-Alvarez, S. Rauch, P. Salaun, J. Schäfer and S. Zimmermann. 2015. COST action TD1407: network on technology-critical elements (NOTICE)—from environmental processes to human health threats. Environ. Sci. Pollut. R. 22: 15188–15194.

Corinaldesi, C., F. Marcellini, E. Nepote, E. Damiani and E. Danovaro R. 2018. Impact of inorganic UV filters contained in sunscreen products on tropical stony corals (*Acropora* spp.). Sci. Total Environ. 637–638: 1279–1285.

Cortés-Gómez, A.A., D. Romero and M. Girondot. 2018a. Carapace asymmetry: A possible biomarker for metal accumulation in adult olive Ridleys marine turtles? Mar. Pollut. Bull. 129: 92–101.

Cortés-Gómez, A.A., A. Tvarijonaviciute, M. Teles, R. Cuenca, G. Fuentes-Mascorro and D. Romero. 2018b. p-Nitrophenyl acetate esterase activity and cortisol as biomarkers of metal pollution in blood of olive ridley turtles (*Lepidochelys olivacea*). Arch. Environ. Con. Tox. 75: 25–36.

Cortés-Gómez, A.A., A. Tvarijonaviciute, M. Girondot, F. Tecles and D. Romero. 2018c. Relationship between plasma biochemistry values and metal concentrations in nesting olive ridley sea turtles. Environ. Sci. Pollut. Res. 25: 36671–36679.

Cox, A., P. Venkatachalam, S. Sahi and N. Sharma. 2016. Silver and titanium dioxide nanoparticle toxicity in plants: A review of current research. Plant Physiol. Bioch. 107: 147–163.

Crosera, M., A. Prodi, M. Mauro, M. Pelin, C. Florio, F. Bellomo, G. Adami, P. Apostoli, G. Palma, M. Bovenzi, M. Campanini and F.L. Filon. 2015. Titanium Dioxide Nanoparticle Penetration into the Skin and Effects on HaCaT Cells. Int. J. Env. Res. Pub. He. 12: 9282–9297.

Cunat, P.-J. 2004. Alloying Elements in stainless steel and other chromium-containing alloys. Euro. Inox. 1–24.

da Nóbrega Alves, R. R., W.L. da Silva Vieira and G.G. Santana. 2008. Reptiles used in traditional folk medicine: Conservation implications. Biodivers. Conserv. 17: 2037–2049.

Dabrunz, A., L. Duester, C. Prasse, F. Seitz, R. Rosenfeldt, C. Schilde, G.E. Schaumann and R. Schulz. 2011. Biological surface coating and molting inhibition as mechanisms of TiO_2 nanoparticle toxicity in *Daphnia magna*. PLOS ONE. 6: e20112.

Dai, Y. and M. Song. 2019. Microstructural evolution and phase transformation of a Ti-5Nb-5Al alloy during annealing treatment. Materials Research. 22: e20190507.

de Almeida Rodrigues, P., R.G. Ferrari, J.V. da Anunciação de Pinho, D.K.A. do Rosário, C.C. de Almeida, T.D. Saint'Pierre, R.A. Hauser-Davis, L.N. Santos and C.A. Conte-Junior. 2022. Baseline titanium levels of three highly consumed invertebrates from an eutrophic estuary in southeastern Brazil. Mar. Pollut. Bull. 183: 114038.

Defense National Stockpile Center. 2008. Strategic and Critical Materials Report to the Congress, Operations under the Strategic and Critical Materials Stock Piling Act during the Period October 2007 through September 2008. United States Department of Defense, USA.

Della Torre, C., F. Buonocore, G. Frenzilli, S. Corsolini, A. Brunelli, P. Guidi, A. Kocan, M. Mariottini, F. Mottola, M. Nigro, K. Pozo, E. Randelli, M.L. Vannuccini, S. Picchietti, M. Santonastaso, V. Scarcelli, S. Focardi, A. Marcomini, L. Rocco, G. Scapigliati and I. Corsi. 2015. Influence of titanium dioxide nanoparticles on 2,3,7,8-tetrachlorodibenzo-p-dioxin bioconcentration and toxicity in the marine fish European sea bass (*Dicentrarchus labrax*). Environ. Pollut. 196: 185–193.

do Amaral, D.F., V. Guerra, K.L. Almeida, L. Signorelli, T.L. Rocha and D. de Melo e Silva. 2022. Titanium dioxide nanoparticles as a risk factor for the health of Neotropical tadpoles: a case study of *Dendropsophus minutus* (Anura: Hylidae). Environ. Sci. Pollut. R. 29: 50515–50529.

Donachie, M.J.Jr. 1988. Titanium: A technical guide. ASM International, Ohio.

Dong, L.-L., H-X. Wang, T. Ding, W. Li and G. Zhang. 2020. Effects of TiO_2 nanoparticles on the life-table parameters, antioxidant indices, and swimming speed of the freshwater rotifer *Brachionus calyciflorus*. J. Exp. Zool. Part A. 333: 230–239.

Doyle, J.J., J.E. Ward and R. Mason. 2015. An examination of the ingestion, bioaccumulation, and depuration of titanium dioxide nanoparticles by the blue mussel (*Mytilus edulis*) and the eastern oyster (*Crassostrea virginica*). Mar. Environ. Res. 110: 45–52.

Dréno, B., A. Alexis, B. Chuberre and M. Marinovich. 2019. Safety of titanium dioxide nanoparticles in cosmetics. J. Eur. Acad. Dermatol. 33: 34–46.

du Preez, M., D. Govender, H. Kylin and H. Bouwman. 2018. Metallic elements in Nile Crocodile eggs from the Kruger National Park, South Africa. Ecotox. Environ. Safe. 148: 930–941.

Duan, Y., H. Liu, J. Zhao, C. Liu, Z. Li, J. Yan, L. Ma, J. Liu, Y. Xie, J. Ruan and F. Hong. 2009. The effects of nano-anatase TiO_2 on the activation of lactate dehydrogenase from rat heart. Biol. Trace Elem. Res. 130: 162–171.

Dulvy, N.K., N. Pacoureau, C.L. Rigby, A.R. Pollom, R.W. Jabado, D.A. Ebert, B. Finucci, C. M. Pollock, J. Cheok, D.H. Derrick, K.B. Herman, C.S. Sherman, W.J. VanderWright, J.M. Lawson, R.H.L. Walls, J.K. Carlson, P. Charvet, K.K. Bineesh, D. Fernando, G.M. Ralph, J.H. Matsushiba, C. Hilton-Taylor, S.V. Fordham and C.A. Simpfendorfer. 2021. Overfishing drives over one-third of all sharks and rays toward a global extinction crisis. Curr. Biol. 31: 4773–4787.e8.

Dutta, P.K., A.K. Ray, V.K. Sharma and F.J. Millero. 2004. Adsorption of arsenate and arsenite on titanium dioxide suspensions. J. Colloid Interf. Sci. 278: 270–275.

Eakin, C.M., H.P.A. Sweatman and R.E. Brainard. 2019. The 2014–2017 global-scale coral bleaching event: insights and impacts. Coral Reefs 38: 539–545.

EFSA, European Food Safety Authority. 2023. Titanium dioxide: E171 no longer considered safe when used as a food additive. Available at: https://www.efsa. europa.eu/en/news/titanium-dioxide-e171-no-longer-considered-safe-when-used-food-additive. (Accessed 15 October 2023).

Egerton, T.A. and I.R. Tooley. 2004. Effect of Changes in TiO_2 dispersion on its measured photocatalytic activity. J. Phys. Chem. B. 108: 5066–5072.

Esther Rubavathi, P., M. Veera Gajendra Babu, B. Bagyalakshmi, L. Venkidu, D. Dhayanithi, N.V. Giridharan and B. Sundarakannan. 2019. Impact of Ba/Ti ratio on the magnetic properties of BaTiO3 ceramics. Vacuum. 159: 374–378.

EU, European Commission. 2020a. Critical Raw Materials for Strategic Technologies and Sectors in the EU, A foresight study. Available at: https:// rmis.jrc.ec.europa.eu/uploads/CRMs_for_Strategic_Technologies_and_ Sectors_in_the_EU_2020.pdf. (Accessed 15 October 2023).

EU, European Commission. 2020b. Opinion on Titanium dioxide (TiO_2) used in cosmetic products that lead to exposure by inhalation. Available at: https:// health.ec.europa.eu/system/files/2021-11/sccs_o_238.pdf. (Accessed 15 October 2023).

Faddah, L.M., N.A. Abdel Baky, N.M. Al-Rasheed and N.M. Al-Rasheed. 2013. Biochemical responses of nanosize titanium dioxide in the heart of rats following administration of idepenone and quercetin. Afr. J. Pharm. Pharmaco. 7: 2639–2651.

Faller, K. and F.H. Froes. 2001. The use of titanium in automobiles: Current technology. JOM. 53: 27.

Fan, W., M. Cui, H. Liu, C. Wang, Z. Shi, C. Tan and X. Yang. 2011. Nano-TiO_2 enhances the toxicity of copper in natural water to Daphnia magna. Environ. Pollut. 159: 729–734.

Fanning, J.C. 2005. Military applications for β titanium alloys. J. Mater. Eng. Perform. 14: 686–690.

Fernández-González, C., G.A. Tarran, N. Schuback, E.M.S. Woodward, J. Arístegui and E. Marañón. 2022. Phytoplankton responses to changing temperature and nutrient availability are consistent across the tropical and subtropical Atlantic. Commun. Biol. 5: 1–13.

Firouzifard, Z., A. Sheikhahmadi and A. Arzinpour. 2016. The effect of additional vitamin E on performance, quality eggs and some blood parameters in Japanese quails during short-term contamination with high levels of Ag-coated titanium dioxide nanoparticles. Anim. Sci. J. 29: 45–56.

Fischer, B.B., F. Pomati and R.I.L. Eggen. 2013. The toxicity of chemical pollutants in dynamic natural systems: The challenge of integrating environmental factors and biological complexity. Sci. Total Environ. 449: 253–259.

Fong, C.R., S.J. Bittick and P. Fong. 2018. Simultaneous synergist, antagonistic and additive interactions between multiple local stressors all degrade algal turf communities on coral reefs. J. Ecol. 106: 1390–1400.

Freiwald, A., J.H. Fossa, A. Grehan, T. Koslow and J.M. Roberts. 2004. Cold-water Coral Reets. UNEP-WCMC, Cambridge, UK.

Freixa, A., V. Acuña, J. Sanchís, M. Farré, D. Barceló and S. Sabater. 2018. Ecotoxicological effects of carbon based nanomaterials in aquatic organisms. Sci. Total Environ. 619-620: 328–337.

Frenzilli, G., M. Bernardeschi, P. Guidi, V. Scarcelli, P. Lucchesi, L. Marsili, M.C. Fossi, A. Brunelli, G. Pojana, A. Marcomini and M. Nigro. 2014. Effects of in vitro exposure to titanium dioxide on DNA integrity of bottlenose dolphin (*Tursiops truncatus*) fibroblasts and leukocytes. Mar. Environ. Res. 100: 68–73.

Froes, F.H. 2015. Titanium: Physical metallurgy, processing, and applications. ASM International, Ohio.

Gaafar, M.S., S.M. Yakout, Y.F. Barakat and W. Sharmoukh. 2022. Electrophoretic deposition of hydroxyapatite/chitosan nanocomposites: The effect of dispersing agents on the coating properties. RSC Adv. 12: 27564–27581.

Gázquez, M.J., J.P. Bolívar, R. Garcia-Tenorio and F. Vaca. 2014. A review of the production cycle of titanium dioxide pigment. Mat. Sci. Appl. 05: 441–458.

Ghosh, M., M. Bandyopadhyay and A. Mukherjee. 2010. Genotoxicity of titanium dioxide (TiO_2) nanoparticles at two trophic levels: plant and human lymphocytes. Chemosphere. 81: 1253–1262.

Gopinath, K.P., N.V. Madhav, A. Krishnan, R. Malolan and G. Rangarajan. 2020. Present applications of titanium dioxide for the photocatalytic removal of pollutants from water: A review. J. Environ. Manage. 270: 110906.

Goreau, T.F., N.I. Goreau and T.J. Goreau. 1979. Corals and Coral Reefs. Sci. Am. 241: 124–137.

Gottschalk, F., T. Sonderer, R.W. Scholz and B. Nowack. 2009. Modeled environmental concentrations of engineered nanomaterials (TiO_2, ZnO, Ag, CNT, Fullerenes) for different regions. Environ. Sci. Technol. 43: 9216–9222.

Gottschalk, F., T. Sonderer, R.W. Scholz and B. Nowack. 2010. Possibilities and limitations of modeling environmental exposure to engineered nanomaterials by probabilistic material flow analysis. Environ. Toxicol. Chem. 29: 1036–1048.

Greenwood, N. N. and A. Earnshaw. 1997. Chemistry of the Elements. Butterworth-Heinemann, Oxford.

Gregor, W. 1791a. Beobachtungen und Versuche über den Menakanit, einen in Cornwall gefundenen magnetischen Sand. Chemische. Annalen. 1:103–119.

Gregor, W. 1791b. Sur le menakanite, espèce de sable attirable par l'aimant, trouvé dans la province de Cornouilles. Observations et Mémoires sur la Physique. 39: 52–160.

Grillitsch, B. and L. Schiesari. 2010. The ecotoxicology of metals in reptiles. pp. 337–448. *In*: Spalding, D.W., G.L. Linder, A. Bishop and S. Krest [eds.]. Ecotoxicology of Amphibians and Reptiles. CRC Press, New York, USA.

Guan, X., W. Shi, S. Zha, J. Rong, W. Su and G. Liu. 2018. Neurotoxic impact of acute TiO$_2$ nanoparticle exposure on a benthic marine bivalve mollusk, *Tegillarca granosa*. Aquat Toxicol. 200: 241–246.

Guan, X., Y. Tang, S. Zha, S. Han, W. Shi, P. Ren, M. Yan, Q. Pan, Y. Hu, J. Fang, J. Zhang and G. Liu. 2019. Exogenous Ca2+ mitigates the toxic effects of TiO$_2$ nanoparticles on phagocytosis, cell viability, and apoptosis in haemocytes of a marine bivalve mollusk, *Tegillarca granosa*. Environ. Pollut. 252: 1764–1771.

Guirlet, E., K. Das and M. Girondot. 2008. Maternal transfer of trace elements in leatherback turtles (*Dermochelys coriacea*) of French Guiana. Aquat Toxicol. 88: 267–276.

Guisbiers, G., S. Mejía-Rosales and F. Leonard Deepak. 2012. Nanomaterial Properties: Size and Shape Dependencies. J. Nanomater. 2012: e180976.

Guitart, C., A. Hernández-del-Valle, J.M. Marín and J. Benedicto. 2012. Tracking temporal trend breaks of anthropogenic change in Mussel Watch (MW) databases. Environ. Sci. Technol. 46: 11515–11523.

Guo, L., W.X. He, P. Zhou and B. Liu. 2020. Research status and de-velopment prospect of titanium and titanium alloy products in China, Hot Work. Technol. 49: 22.

Hale, R.C., M.E. Seeley, M.J. La Guardia, L. Mai and E.Y. Zeng. 2020. A global perspective on microplastics. J. Geophys. Res-Oceans 125: e2018JC014719.

Hammond, S.A., A. Carew and C. Helbing. 2013. Evaluation of the effects of titanium dioxide nanoparticles on cultured *Rana catesbeiana* tailfin tissue. Frontiers in Genetics. 4: 251.

Hao, L., Z. Wang and B. Xing. 2009. Effect of sub-acute exposure to TiO$_2$ nanoparticles on oxidative stress and histopathological changes in Juvenile Carp (*Cyprinus carpio*). J. Environ. Sci. 21: 1459–1466.

Hartmann, N.B., S. Legros, F. Von der Kammer, T. Hofmann and A. Baun. 2012. The potential of TiO$_2$ nanoparticles as carriers for cadmium uptake in *Lumbriculus variegatus* and *Daphnia magna*. Aquat. Toxicol. 118–119: 1–8.

Hauser-Davis, R.A., F. Monteiro, D. Cardoso, D. and S. Siciliano. 2020. Titanium as a contaminant of emerging concern in the aquatic environment and the current knowledge gap regarding seabird contamination. Ornithologia 11: 7–15.

Hauser-Davis, R.A., R.C.C. Rocha, T.D. Saint'Pierre and D.H. Adams. 2021. Metal concentrations and metallothionein metal detoxification in blue sharks, *Prionace glauca* L. from the Western North Atlantic Ocean. J. Trace Elem. Med. Bio. 68: 126813.

Hauser-Davis, R. A. 2023. Subcellular cadmium, lead and mercury partitioning assessments in aquatic organisms as a tool for assessing actual toxicity and trophic transfer. *In*: R.A. Hauser-Davis, N.S. Quinete and L.S. Lemos [eds.]. Lead, Mercury and Cadmium in the Aquatic Environment. CRC Press, Boca Raton.

Heinlaan, M., A. Ivask, I. Blinova, H.C. Dubourguier and A. Kahru. 2008. Toxicity of nanosized and bulk ZnO, CuO and TiO_2 to bacteria *Vibrio fischeri* and crustaceans *Daphnia magna* and *Thamnocephalus platyurus*. Chemosphere. 71: 1308–1316.

Holsbeek, L., U. Siebert and C.R. Joiris. 1998. Heavy metals in dolphins stranded on the French Atlantic coast. Sci. Total Environ. 217: 241–249.

Hoppe, C.J.M., K.K.E. Wolf, N. Schuback, P.D. Tortell and B. Rost. 2018. Compensation of ocean acidification effects in Arctic phytoplankton assemblages. Nat. Clim. Change. 8: 529–533.

Hosseini, M., S.H. Rahmanpour and A. Mashinchian Moradi. 2015. Heavy metal ions on titanium dioxide nano-particle: Biomagnification in an experimental aquatic food chain. International Journal of Marine Science and Environment. 5:23–29.

Hou, J., L. Wang, C. Wang, S. Zhang, H. Liu, S. Li and X. Wang. 2019. Toxicity and mechanisms of action of titanium dioxide nanoparticles in living organisms. J. Environ. Sci. 75: 40–53.

Hristozov, D.R., S. Gottardo, S. Critto and A. Marcomini. 2012. Risk assessment of engineered nanomaterials: A review of available data and approaches from a regulatory perspective. Nanotoxicology 6: 880–898.

Huang, X., Z. Liu, Z. Xie, S. Dupont, W. Huang, F. Wu, H. Kong, L. Liu, Y. Sui, D. Lin, W. Lu, M. Hu and Y. Wang. 2018. Oxidative stress induced by titanium dioxide nanoparticles increases under seawater acidification in the thick shell mussel *Mytilus coruscus*. Mar. Environ. Res. 137: 49–59.

Hunt, R.J. and V.F. Matveev. 2005. The effects of nutrients and zooplankton community structure on phytoplankton growth in a subtropical Australian reservoir: An enclosure study. Limnologica. 35: 90–101.

Hunter, M.A. 1910. Metallic Titanium. J. Am. Chem. Soc. 32: 330–336.

IARC, International Agency for Research on Cancer. 2006. Titanium dioxide (IARC Group 2B), Summary of data reported. IARC Monographs on the Evaluation of Carcinogenic Risks to Humans. 93: 272–276.

Ingelmann, C.-J., M. Witzig, J. Möhring, M. Schollenberger, I. Kühn and M. Rodehutscord. 2018. Effect of supplemental phytase and xylanase in wheat-based diets on prececal phosphorus digestibility and phytate degradation in young turkeys. Poultry Sci. 97: 2011–2020.

Ishii, M., T. Oda and M. Kaneko. 2003. Titanium and its alloys as key materials for corrosion protection engineering. Nippon Steel Technical Report No. 87. 45–50.

Iswarya, V., V. Sharma, N. Chandrasekaran and A. Mukherjee. 2017. Impact of tetracycline on the toxic effects of titanium dioxide (TiO_2) nanoparticles towards the freshwater algal species, *Scenedesmus obliquus*. Aquat Toxicol. 193: 168–177.

Jafari, S., B. Mahyad, H. Hashemzadeh, S. Janfaza, T. Gholikhani and L. Tayebi. 2023. Biomedical applications of TiO$_2$ nanostructures: Recent Advances. Int. J. Nanomed. 15: 3447–3470.

Jagnaux, R. 1891. Histoire de la chimie. Baudry et cie, Paris.

Jasper, A. 2020. Architecture and Anthropology. Routledge, Taylor & Francis, Oxon.

Jiang, J., G. Oberdörster and P. Biswas. 2009. Characterization of size, surface charge, and agglomeration state of nanoparticle dispersions for toxicological studies. J. Nanopart. Res. 11: 77–89.

Jiménez-Tototzintle, M., I. Oller, A. Hernández-Ramírez, S. Malato and M.I. Maldonado. 2015. Remediation of agro-food industry effluents by biotreatment combined with supported TiO$_2$/H$_2$O$_2$ solar photocatalysis. Chem. Eng. J. 273: 205–213.

Jomini, S., H. Clivot, P. Bauda and C. Pagnout. 2015. Impact of manufactured TiO$_2$ nanoparticles on planktonic and sessile bacterial communities. Environ. Pollut. 202: 196–204.

Jovanović, B. and H.M. Guzmán. 2014. Effects of titanium dioxide (TiO$_2$) nanoparticles on caribbean reef-building coral (*Montastraea faveolata*). Environ. Toxicol. Chem. 33: 1346–1353.

Kalman, J., K.B. Paul, F.R. Khan, V. Stone and T.F. Fernandes. 2015. Characterisation of bioaccumulation dynamics of three differently coated silver nanoparticles and aqueous silver in a simple freshwater food chain. Environ. Chem. 12: 662–672.

Kannegiesser, M. 2008. Value chain management in the chemical industry. contributions to management science. Physica-Verlag HD, Heidelberg.

Keller, A.A., H. Wang, D. Zhou, H.S. Lenihan, G. Cherr, B.J. Cardinale, R. Miller and Z. Ji. 2010. Stability and aggregation of metal oxide nanoparticles in natural aqueous matrices. Environ. Sci. Technol. 44: 1962–1967.

Keller, A.A., S. McFerran, A. Lazareva and S. Suh. 2013. Global life cycle releases of engineered nanomaterials. J. Nanopart. Res. 15: 1692.

Khan, S., Mu. Naushad, M. Govarthanan, J. Iqbal and S.M. Alfadul. 2022. Emerging contaminants of high concern for the environment: Current trends and future research. Environ. Res. 207: 112609.

Klaproth, M.H. 1801. Analytical essays towards gromoting the chemical knowledge of mineral substances. Cadell and Davies, London.

Klein, R., M. Bartel-Steinbach, J. Koschorreck, M. Paulus, K. Tarricone, D. Teubner, G. Wagner, T. Weimann and M. Veith. 2012. Standardization of egg collection from aquatic birds for biomonitoring - A critical review. Environ. Sci. Technol. 46: 5273–5284.

Kleinow, K., J. Baker, J. Nichols, F. Gobas, T. Parkerton, D. Muir, G. Monteverdi and P. Mastrodone. 1999. Exposure, uptake, and disposition of chemicals in reproductive and developmental stages of selected oviparous vertebrates. pp. 9–111 *In:* Di Giulo, R.T. and D.E. Tillitt [eds.]. Reproductive and developmental effects of contaminants in oviparous vertebrates. SETAC Press, Pensacola, FL.

Kong, H., F. Wu, X. Jiang, T. Wang, M. Hu, J. Chen, W. Huang, Y. Bao and Y. Wang. 2019. Nano-TiO$_2$ impairs digestive enzyme activities of marine mussels under ocean acidification. Chemosphere. 237: 124561.

Krebs, R.E. 2006. The History and Use of Our Earth's Chemical Elements: A Reference Guide. Greenwood Publishing Group, London.

Krishna, R., A.D. Dhass, A. Arya, R. Prasad and I. Colak. 2023. An assessment of the strategies for the energy-critical elements necessary for the development of sustainable energy sources. Environ. Sci. Pollut. Res. 1–22.

Labille, J., C. Harns, J.Y. Bottero and J. Brant. 2015. Heteroaggregation of titanium dioxide nanoparticles with natural clay colloids. Environ. Sci. Technol. 49: 6608–6616.

Labille, J., D. Slomberg, R. Catalano, S. Robert, M.L. Apers-Tremelo, J.L. Boudenne, T. Manasfi and O. Radakovitch. 2020. Assessing UV filter inputs into beach waters during recreational activity: A field study of three French Mediterranean beaches from consumer survey to water analysis. Sci. Total Environ. 706: 136010.

Laitano, M.V. and A.V. Fernández-Gimenez. 2016. Are mussels always the best bioindicators? comparative study on biochemical responses of three marine invertebrate species to chronic port pollution. B. Environ. Contam. Tox. 97: 50–55.

Lakshmanan, V.I., A. Bhowmick an M. Abdul Halim. 2014. Titanium dioxide: production, properties and applications. pp. 75–130. *In:* Brown, J. [ed.]. Titanium dioxide. Nova Science Publishers, New York.

Larue, C., G. Veronesi, A.M. Flank, S. Surble, N. Herlin-Boime and M. Carrière. 2012. Comparative uptake and impact of TiO$_2$ nanoparticles in wheat and rapeseed. J. Toxicol. Env. Heal. A. 75: 722–734.

Lee, C.-C., Y.-H. Lin, W.-C. Hou, M.-H. Li and J.-W. Chang. 2020. Exposure to ZnO/TiO$_2$ nanoparticles affects health outcomes in cosmetics salesclerks. Int. J. Env. Res. Pub. He. 17: 6088.

Li, L., M. Sillanpää, M. Tuominen, K. Lounatmaa and E. Schultz. 2013. Behavior of titanium dioxide nanoparticles in Lemna minor growth test conditions. Ecotox. Environ. Safe. 88: 89–94.

Li, M., Y. Zhang, S. Feng, X. Zhang, Y. Xi and X. Xiang. 2022. Bioaccumulation and biomagnification effects of nano-TiO$_2$ in the aquatic food chain. Ecotoxicology. 31: 1023–1034.

Lindenschmidt, R.C., K.E. Driscoll, M.A. Perkins, J.M. Higgins, J.K. Maurer and K.A. Belfiore. 1990. The comparison of a fibrogenic and two nonfibrogenic dusts by bronchoalveolar lavage. Toxicol. Appl. Pharm. 102: 268–281.

Linnik, P.N. and V.A. Zhezherya. 2015. Titanium in natural surface waters: The content and coexisting forms. Russ. J. Gen. Chem. 85: 2908–2920.

Liriano-Jorge, C.F., U. Sohmen, A. Özkan, H. Gulyas and R. Otterpohl. 2014. TiO$_2$ photocatalyst nanoparticle separation: flocculation in different matrices and use of powdered activated carbon as a precoat in low-cost fabric filtration. Adv. Mater. Sci. Eng. 2014: e602495.

Litchman, E., P. Tezanos Pinto, K.F. Edwards, C.A. Klausmeier, C.T. Kremer and M.K. Thomas. 2015. Global biogeochemical impacts of phytoplankton: A trait-based perspective. J. Ecol. 103: 1384–1396.

Liu, X., P.K. Chu and C. Ding. 2004. Surface modification of titanium, titanium alloys, and related materials for biomedical applications. Mat. Sci. Eng. R. 47: 49–121.

Liu, M., X. Pan, Y. Gan, M. Gao, X. Li, Z. Liu, X. Ma, M. Geng, X. Meng, N. Ma and J. Li. 2023. Titanium carbide MXene quantum dots-modified hydroxyapatite hollow microspheres as pH/near-infrared dual-response drug carriers. Langmuir. 39: 13325–13334.

Livraghi, S., M. Chiesa, M.C. Paganini and E. Giamello. 2011. On the nature of reduced states in titanium dioxide as monitored by electron paramagnetic resonance. I: The Anatase Case. J. Phys. Chem. C. 115: 25413–25421.

Loosli, F., P. Le Coustumer and S. Stoll. 2013. TiO_2 nanoparticles aggregation and disaggregation in presence of alginate and Suwannee River humic acids. pH and concentration effects on nanoparticle stability. Water Res. 47: 6052–6063.

Łosiewicz, B., A. Stróż, P. Osak, J. Maszybrocka, A. Gerle, K. Dudek, K. Balin, K., D. Łukowiec, M. Gawlikowski and S. Bogunia. 2021. Production, characterization and application of oxide nanotubes on Ti–6Al–7Nb alloy as a potential drug carrier. Mater. 14: 6142.

Lu, J., S. Tian, X. Lv, Z. Chen, B. Chen, X. Zhu and Z. Cai. 2018. TiO_2 nanoparticles in the marine environment: Impact on the toxicity of phenanthrene and Cd^{2+} to marine zooplankton *Artemia salina*. Sci. Total Environ. 615: 375–380.

Lu, G., A. Zhu, H. Fang, Y. Dong and W.X. Wang. 2019. Establishing baseline trace metals in marine bivalves in China and worldwide: Meta-analysis and modeling approach. Sci. Total Environ. 669: 746–753.

Luo, H., Y. Xiang, Y. Li, Y. Zhao and X. Pan. 2020. Weathering alters surface characteristic of TiO_2-pigmented microplastics and particle size distribution of TiO_2 released into water. Sci. Total Environ. 729: 139083.

Lv, X., J. Tao, B. Chen and X. Zhu. 2016. Roles of temperature and flow velocity on the mobility of nano-sized titanium dioxide in natural waters. Sci. Total Environ. 565: 849–856.

Magdolenova, Z., A. Collins, A. Kumar, A. Dhawan, V. Stone and M. Dusinska. 2014. Mechanisms of genotoxicity. A review of *in vitro* and *in vivo* studies with engineered nanoparticles. Nanotoxicology. 8: 233–278.

Matouke, M.M. and M. Mustapha. 2018. Bioaccumulation and physiological effects of copepods sp. (*Eucyclop* sp.) fed *Chlorella ellipsoides* exposed to titanium dioxide (TiO_2) nanoparticles and lead (Pb^{2+}). Aquat Toxicol. 198: 30–39.

Matsuo, S., Y. Anraku, S. Yamada, T. Honjo, T. Matsuo and H. Wakita. 2001. Effects of photocatalytic reactions on marine plankton: Titanium dioxide powder as catalyst on the body surface. J. Environ. Sci. Heal. A. 36: 1419–1425.

Maul, G.A. and I.W. Duedall. 2019. Demography of coastal populations. pp. 692–700. *In:* Finkl, C.W. and C. Makowski [eds.]. Encyclopedia of Coastal Science. Springer, Cham.

McIntyre, T. and M.J. Whiting. 2012. Increased metal concentrations in giant sungazer lizards (*Smaug giganteus*) from mining areas in South Africa. Arch. Environ. Con. Tox. 63: 574–585.

McIntyre, L., P.B. Patterson, L.C. Anderson and C.L. Mah. 2016. Household food insecurity in canada: Problem definition and potential solutions in the public policy domain. Can. Public Pol. 42: 83–93.

Meinhold, G. 2010. Rutile and its applications in earth sciences. Earth-Sci. Rev. 102: 1–28.

Miller, R.J., S. Bennett, A.A. Keller, S. Pease and H.S. Lenihan. 2012. TiO$_2$ Nanoparticles are phototoxic to marine phytoplankton. PLOS ONE. 7: e30321.

Miller, I.B., S. Pawlowski, M.Y. Kellermann, M. Petersen-Thiery, M. Moeller, S. Nietzer and P.J. Schupp. 2021. Toxic effects of UV filters from sunscreens on coral reefs revisited: regulatory aspects for "reef safe" products. Environmental Sciences Europe. 33: 74.

Mishra, S.P. 2020. Significance of fish nutrients for human health. Int. J. Fish. Aquat. Res. 5: 47–49.

Mohseni, E., E. Zalnezhad, A.R. Bushroa, Abdel Magid Hamouda, B.T. Goh and G.H. Yoon. 2015. Ti/TiN/HA coating on Ti–6Al–4V for biomedical applications. Ceram. Internat. 41: 14447–14457.

Monteiro, F., L.S. Lemos, J.F. de Moura, R.C.C. Rocha, I. Moreira, A.P. Di Beneditto, H.A. Kehrig, I. C.A.C. Bordon, S. Siciliano, T.D. Saint'Pierre and R.A. Hauser-Davis. 2019. Subcellular metal distributions and metallothionein associations in rough-toothed dolphins (*Steno bredanensis*) from Southeastern Brazil. Mar. Pollut. Bull. 146: 263–273.

Monteiro, F., L.S. Lemos, J.F. de Moura, R.C.C. Rocha, I. Moreira, A.P.M. Di Beneditto, H.A. Kehrig, I.C. Bordon, S. Siciliano, T.D. Saint'Pierre and R.A. Hauser-Davis. 2020. Total and subcellular Ti distribution and detoxification processes in *Pontoporia blainvillei* and *Steno bredanensis* dolphins from Southeastern Brazil. Mar. Pollut. Bull. 153: 110975.

Nagle, R.D., C.L. Rowe and J.D. Congdon. 2001. Accumulation and selective maternal transfer of contaminants in the turtle trachemys scripta associated with coal ash deposition. Arch. Environ. Con. Tox. 40: 531–536.

Nations, S., M. Wages, J.E. Cañas, J. Maul, C. Theodorakis and G.P. Cobb. 2011. Acute effects of Fe$_2$O$_3$, TiO$_2$, ZnO and CuO nanomaterials on *Xenopus laevis*. Chemosphere. 83: 1053–1061.

Navidpour, A.H., S. Abbasi, D. Li, A. Mojiri and J.L. Zhou. 2023. Investigation of advanced oxidation process in the presence of tio$_2$ semiconductor as photocatalyst: Property, Principle, Kinetic Analysis, and Photocatalytic Activity. Catalysts. 13: 232.

Nigro, M., M. Bernardeschi, D. Costagliola, C. Della Torre, G. Frenzilli, P. Guidi, P. Lucchesi, F. Mottola, M. Santonastaso, V. Scarcelli, F. Monaci, I. Corsi, V. Stingo and L. Rocco. 2015. n-TiO$_2$ and CdCl$_2$ co-exposure to titanium dioxide nanoparticles and cadmium: Genomic, DNA and chromosomal damage evaluation in the marine fish European sea bass (*Dicentrarchus labrax*). Aquat Toxicol. 168: 72–77.

Nuss, P. and G.A. Blengini. 2018. Towards better monitoring of technology critical elements in Europe: Coupling of natural and anthropogenic cycles. Sci. Total Environ. 613-614: 569–578.

Nyamukamba, P., O. Okoh, H. Mungondori, R. Taziwa and S. Zinya. 2018. Synthetic methods for titanium dioxide nanoparticles: A review. pp. 151–157. *In:* Yang, D. [ed.] Titanium Dioxide - Material for a Sustainable Environment. IntechOpen, London.

Oliveira, C.C.V., L. Ferrão, V. Gallego, C. Mieiro, I.B. Oliveira, A. Carvalhais, M. Pachedo and E. Cabrita. 2023. Exposure to silver and titanium dioxide nanoparticles at supra-environmental concentrations decreased sperm motility and affected spermatozoa subpopulations in gilthead seabream, *Sparus aurata*. Fish Physiol. Biochem. 1–12.

Ophus, E.M., L. Rode, B. Gylseth, D.G. Nicholson and K. Saeed. 1979. Analysis of titanium pigments in human lung tissue. Scand. J. Work Env. Hea. 5: 290–296.

Park, E.J., J. Yoon, K. Choi, J. Yi and K. Park. 2009. Induction of chronic inflammation in mice treated with titanium dioxide nanoparticles by intratracheal instillation. Toxicology. 260: 37–46.

Patel, P. 2016. Toxic Cosmetics: Lead in Lipstick. Bioclinic Naturals.

Pitre, S.P., T.P. Yoon and J.C. Scaiano. 2017. Titanium dioxide visible light photocatalysis: surface association enables photocatalysis with visible light irradiation. Chem. Commun. 53: 4335–4338.

Poynton, H.C. and W.E. Robinson. 2018. Contaminants of emerging concern, whith a emphasis on nanomaterials and pharmaceuticals. pp. 291–311. *In:* B. Török and T. Dransfield [eds.]. Green Chemistry: an inclusive approach. Elsevier, Amsterdam.

Qiu, G. and Y. Guo. 2022. Current situation and development trend of titanium metal industry in China. Int. J. Min. Met. Mater. 29: 599–610.

Raines, A.L., R. Olivares-Navarrete, M. Wieland, D.L. Cochran, Z. Schwartz and B.D. Boyan. 2010. Regulation of angiogenesis during osseointegration by titanium surface microstructure and energy. Biomaterials. 31: 4909–4917.

Ramachandran, P., C. Yew Lee, R.-A. Doong, C. Ein Oon, N.T.K. Thanh and H. Ling Lee. 2020. A titanium dioxide/nitrogen-doped graphene quantum dot nanocomposite to mitigate cytotoxicity: Synthesis, characterisation, and cell viability evaluation. RSC Adv. 10: 21795–21805.

Ramsden, C.S., T.B. Henry and R.D. Handy. 2013. Sub-lethal effects of titanium dioxide nanoparticles on the physiology and reproduction of zebrafish. Aquat Toxicol. 126: 404–413.

Reef Resilience Network. 2022. Bleaching Impacts, Reef Resilience. Available at: https://reefresilience.org/stressors/bleaching/bleaching-impacts/. (Accessed 15 October 2023).

Reeves, J.F., S.J. Davies, N.J.F. Dodd and A.N. Jha. 2008. Hydroxyl radicals (OH) are associated with titanium dioxide (TiO_2) nanoparticle-induced cytotoxicity and oxidative DNA damage in fish cells. Mutat. Res-Fund. Mol. M. 640: 113–122.

Rhen, T. and J.A. Cidlowski. 2005. Antiinflammatory action of glucocorticoids—new mechanisms for old drugs. New Engl. J. Med. 353: 1711–1723.

Rich, C.N. and L.G. Talent. 2009. Soil ingestion may be an important route for the uptake of contaminants by some reptiles. Environ. Toxicol. Chem. 28: 311–315.

Ripple, W.J., J.A. Estes, O.J. Schmitz, V. Constant, M.J. Kaylor, A. Lenz, J.A. Motley, K.E. Self, D.S. Taylor and C. Wolf. 2016. What is a trophic cascade? Trends Ecol. Evol. 31: 842–849.

Robichaud, C.O., A.E. Uyar, M.R. Darby, L.G. Zucker and M.R. Wiesner. 2009. Estimates of upper bounds and trends in Nano-TiO$_2$ production as a basis for exposure assessment. Environ. Sci. Technol. 43: 4227–4233.

Rocco, L., M. Santonastaso, M. Nigro, F. Mottola, D. Costagliola, M. Bernardeschi, P. Guidi, P. Lucchesi, V. Scarcelli, I. Corsi, V. Stingo and G. Frenzilli. 2015. Genomic and chromosomal damage in the marine mussel *Mytilus galloprovincialis*: Effects of the combined exposure to titanium dioxide nanoparticles and cadmium chloride. Mar. Environ. Res. 111: 144–148.

Rocha, S.S.D., G.L. Adabo, G.E.P. Henriques and M.A.D.A. Nóbilo. 2006. Vickers hardness of cast commercially pure titanium and Ti-6Al-4V alloy submitted to heat treatments. Brazilian Dental Journal. 17: 126–129.

Rodrigues, P. de A., R.G. Ferrari, D.K.A. Rosário, C.C. de Almeida, T.D. Saint'Pierre, R.A. Hauser-Davis, L.N. dos Santos and C.A. Conte-Junior. 2022. Toxic metal and metalloid contamination in seafood from an eutrophic Brazilian estuary and associated public health risks. Mar. Pollut. Bull. 185: 114367.

Romero-Freire, A., J. Santos-Echeandía, P. Neira and A. Cobelo-García. 2019. Less-studied technology-critical elements (Nb, Ta, Ga, In, Ge, Te) in the marine environment: Review on their concentrations in water and organisms. Frontiers in Marine Science. 6.

Ropers, M.H., H. Terrisse, M.Mercier-Bonin and B. Humbert. 2017. Titanium dioxide as food additive. pp. 3–19. *In*: Janus, M. [ed.]. Application of Titanium Dioxide. IntechOpen, Croatia.

RSC, Royal Society of Chemistry. 2022. Periodic Table. Available at: https://www. rsc.org/periodic-table. (Accessed 15 October 2023).

RSC, Royal Society of Chemistry. 2023. Titanium. Available at: https://www.rsc. org/periodic-table/element/22/titanium. (Accessed 19 October 2023).

Rosenfeldt, R.R., F. Seitz, R. Schulz and M. Bundschuh. 2014. Heavy metal uptake and toxicity in the presence of titanium dioxide nanoparticles: A factorial approach using *Daphnia magna*. Environ. Sci. Technol. 48: 6965–6972.

Roza, G. 2008. Titanium. The Rosen Publishing Group, Inc, New York.

Saggioro, E.M., A.S. Oliveira, T. Pavesi, C.G. Maia, L.F.V. Ferreira and J.C. Moreira. 2011. Use of titanium dioxide photocatalysis on the remediation of model textile wastewaters containing azo dyes. Molecules. 16: 10370–10386.

Saggioro, E., A. Oliveira, T. Pavesi and J. Moreira. 2014. Effect of activated carbon and titanium dioxide on the remediation of an indigoid dye in model waters. Rev. Chim-Bucharest. 65: 237–241.

Santos Filho, R., T. Vicari, S.A. Santos, K. Felisbino, N. Mattoso, B.F. Sant'Anna-Santos, M.M. Cestari, M.M. and D.M. Leme. 2019. Genotoxicity of titanium dioxide nanoparticles and triggering of defense mechanisms in *Allium cepa*. Genet. Mol. Biol. 42: 425–435.

Sarraf, M., A. Dabbagh, B. Abdul Razak, B. Nasiri-Tabrizi, H.R.M. Hosseini, S. Saber-Samandari, N.H. Abu Kasim, L.K. Yean and N.L. Sukiman. 2018. Silver oxide nanoparticles-decorated tantala nanotubes for enhanced antibacterial activity and osseointegration of Ti6Al4V. Mater. Design. 154: 28–40.

Sarraf, M., E. Rezvani Ghomi, S. Alipour, S. Ramakrishna and N. Liana Sukiman. 2022. A state-of-the-art review of the fabrication and characteristics of titanium and its alloys for biomedical applications. Bio-Design and Manufacturing. 5: 371–395.

Schrand, A.M., M.F. Rahman, S.M. Hussain, J.J. Schlager, D.A. Smith and A.F. Syed. 2010. Metal-based nanoparticles and their toxicity assessment. WIREs Nanomed. Nanobi. 2: 544–568.

Seagle, S.R. 2019. Titanium processing. Encyclopedia Britannica, 17 May. Available at: https://www.britannica.com/technology/titanium-processing. (Accessed 25 December 2022).

Senzui, M., T. Tamura, K. Miura, Y. Ikarashi, Y. Watanabe and M. Fujii. 2010. Study on penetration of titanium dioxide (TiO_2) nanoparticles into intact and damaged skin *in vitro*. J. Toxicol. Sci. 35: 107–113.

Shahid, M., E. Ferrand, E. Schreck and C. Dumat. 2012. Behavior and impact of zirconium in the soil–plant system: Plant uptake and phytotoxicity. pp. 107–127. *In*: Whitacre, D.M. [ed.]. Reviews of Environmental Contamination and Toxicology Springer, New York.

Shaw, B.J., C.S. Ramsden, A. Turner and R.D. Handy. 2013. A simplified method for determining titanium from TiO_2 nanoparticles in fish tissue with a concomitant multi-element analysis. Chemosphere. 92: 1136–1144.

Sheng, L., L. Wang, M. Su, X. Zhao, R. Hu, X. Yu, J. Hong, D. Liu, B. Xu, Y. Zhu, H. Wang and F. Hong. 2016. Mechanism of TiO_2 nanoparticle-induced neurotoxicity in zebrafish (*Danio rerio*). Environ. Toxicol. 31: 163–175.

Shi, H., R. Magaye, V. Castranova and J. Zhao. 2013. Titanium dioxide nanoparticles: a review of current toxicological data. Part. Fibre Toxicol. 10: 15.

Shi, W., Y. Han, C. Guo, W. Su, X. Zhao, S. Zha, Y. Wang and G. Liu. 2019. Ocean acidification increases the accumulation of titanium dioxide nanoparticles ($nTiO_2$) in edible bivalve mollusks and poses a potential threat to seafood safety. Sci. Rep-UK. 9: 3516.

Sigman, D.M. and M.P. Hain. 2012. The biological productivity of the ocean. Nat. Educ. Knowledge. 3: 1–16.

Sigworth, G.K. 2008. The modification of Al-Si casting alloys: important practical and theoretical aspects. Int. J. Metalca. St. 2: 19–40.

Sirés, I. and E. Brillas. 2012. Remediation of water pollution caused by pharmaceutical residues based on electrochemical separation and degradation technologies: A review. Environ. Int. 40: 212–229.

Skocaj, M., M. Filipic, J. Petkovic and S. Novak. 2011. Titanium dioxide in our everyday life; is it safe? Radiol. Oncol. 45: 227–247.

Sohn, H.-S. 2020. Production technology of titanium by Kroll process. Resour. Recy. 29: 3–14.

Solomon, B.D. 2023. Millennium ecosystem assessment. pp. 352–353. *In*: Haddad, B.M. and B.D. Solomon [eds.]. Dictionary of Ecological Economics, Terms for the New Millennium. Edward Elgar Publishing Limited, Cheltenham, UK.

Song, U., M. Shin, G. Lee, J. Roh, Y. Kim and E.J. Lee. 2013. Functional analysis of TiO_2 nanoparticle toxicity in three plant species. Biol. Trace Elem. Res. 155: 93–103.

Soto, M., M.P. Ireland and I. Marigómez. 2000. Changes in mussel biometry on exposure to metals: Implications in estimation of metal bioavailability in 'Mussel-Watch' programmes. Sci. Total Environ. 247: 175–187.

Souza, J.C.M., M.B. Sordi, M. Kanazawa, S. Ravindran, B. Henriques, F.S. Silva, C. Aparicio and L.F. Cooper. 2019. Nano-scale modification of titanium implant surfaces to enhance osseointegration. Acta Biomater. 94: 112–131.

Stasinakis, A.S. 2008. Use of selected advanced oxidation processes (AOPs) for wastewater treatment—A mini review. Global Nest J. 10: 376–385.

Statista. 2022. Market size of titanium worldwide in 2021 and 2022, with a forecast for 2023 and 2030. Available at: https://www.statista.com/statistics/1318587/market-size-of-titanium-worldwide/. (Accessed 15 October 2023).

Sun, H., X. Zhang, Q. Niu, Y. Chen and J.C. Crittenden. 2007. Enhanced accumulation of arsenate in carp in the presence of titanium dioxide nanoparticles. Water Air Soil Poll. 178: 245–254.

Supriya, R.A., S. Sureshkannan, K. Porteen, B.S.M. Ronald, K.G. Tirumurugaan, A. Uma and A. Sangeetha. 2020. Investigation of heavy metal concentrations in sea food from three selected landing centers of Chennai coast. Int. J. Chem. Stud. 8: 08–14.

Susanto, G.N. 2021. Crustacea: The increasing economic importance of crustaceans to humans. *In*: Ranz, R.E.R. [ed.]. Arthropods—Are They Beneficial for Mankind? IntechOpen, London.

Suzuki, K., J. Noda, M. Yanagisawa, I. Kawazu, K. Sera, D. Fukui, M. Asakawa and H. Yokota. 2012. Particle-induced X-ray emission analysis of elements in plasma from wild and captive sea turtles (*Eretmochelys imbricata*, *Chelonia mydas*, and *Caretta caretta*) in Okinawa, Japan. Biol. Trace Elem. Res. 148: 302–308.

Szota, M., A. Łukaszewicz and K. Machnik. 2020. The possibility to control the thickness of the oxide layer on the titanium Grade 2 by mechanical activation and heat treatment. Journal of Achievements in Materials and Manufacturing Engineering. 2: 70–77.

Takeda, K., K. Suzuki, A. Ishihara, M. Kubo-Irie, R. Fujimoto, M. Tabata, S. Oshio, Y. Nihei, T. Ihara and M. Sugamata. 2009. Nanoparticles transferred from pregnant mice to their offspring can damage the genital and cranial nerve systems. J. Health Sci. 55: 95–102.

Takeda, O., T. Ouchi and T.H. Okabe. 2020. Recent progress in titanium extraction and recycling. Metall. Mater. Trans. B. 51: 1315–1328.

Tan, C., W.H. Fan and W.X. Wang. 2012. Role of titanium dioxide nanoparticles in the elevated uptake and retention of cadmium and zinc in *Daphnia magna*. Am. Chem. S. 46: 469–476.

Tassinari, R., F. Cubadda, G. Moracci, F. Aureli, M. D'Amato, M. Valeri, B. De Berardis, A. Raggi, A. Mantovani, D. Passeri, M. Rossi and F. Maranghi. 2014. Oral, short-term exposure to titanium dioxide nanoparticles in Sprague-Dawley rat: Focus on reproductive and endocrine systems and spleen. Nanotoxicology. 8: 654–662.

Tay, C.Y., W. Fang, M.I. Setyawati, S.L. Chia, K.S. Tan, C.H.L. Hong and D.T. Leong. 2014. Nano-hydroxyapatite and nano-titanium dioxide exhibit different subcellular distribution and apoptotic profile in human oral epithelium. ACS Applied Materials & Interfaces. 6: 6248–6256.

Teng, F., M. Li, C. Gao, G. Zhang, P. Zhang, Y. Wang, L. Chen and E. Xie. 2014. Preparation of black TiO_2 by hydrogen plasma assisted chemical vapor deposition and its photocatalytic activity. Appl. Catal. B-Environ. 148-149: 339–343.

Thabet, A.F., O.A. Galal, M.F.M. El-Samahy and M. Tuda. 2019. Higher toxicity of nano-scale TiO_2 and dose-dependent genotoxicity of nano-scale SiO2 on the cytology and seedling development of broad bean Vicia faba. SN Applied Sciences. 1: 956.

Thesiya, D., J. Dave, A. Rajurkar and V. Prajapati. 2015. Study of influence of EDM process parameters during machining of TI-6AL-4V. J. Manuf. Tech. Res. 7: 53.

Thiagarajan, V., L. Natarajan, R. Seenivasan, N. Chandrasekaran and A. Mukherjee. 2019a. Tetracycline affects the toxicity of P25 n-TiO_2 towards marine microalgae *Chlorella* sp. Environ. Res. 179: 108808.

Thiagarajan, V., V.P.A.J. Iswarya, R. Seenivasan, N. Chandrasekaran and A. Mukherjee. 2019b. Influence of differently functionalized polystyrene microplastics on the toxic effects of P25 TiO_2 NPs towards marine algae *Chlorella* sp. Aquat Toxicol. 207: 208–216.

Thiagarajan, V., R. Seenivasan, D. Jenkins, N. Chandrasekaran and A. Mukherjee. 2020. Combined effects of nano-TiO_2 and hexavalent chromium towards marine crustacean *Artemia salina*. Aquat Toxicol. 225: 105541.

Tsukano, K., K. Suzuki, J. Noda, M. Yanagisawa, K. Kameda, K. Sera, Y. Nishi, T. Shimamori, Y. Morimoto, H. Yokota and M. Asakawa. 2017. Plasma lead, silicon and titanium concentrations are considerably higher in green sea turtle from the suburban coast than in those from the rural coast in Okinawa, Japan. J. Vet. Med. Sci. 79: 2043–2047.

USEPA, US Environmental Protection Agency. 2010. Emerging Contaminants-Nanomaterials. Solid Waste and EPA 505-F-10-008. United States Environmental Protection Agency, Washington.

USEPA, US Environmental Protection Agency. 2023. Reviewing New Chemicals under the Toxic Substances Control Act (TSCA). Available at: https://www.epa.gov/reviewing-new-chemicals-under-toxic-substances-control-act-tsca/control-nanoscale-materials-under. (Accessed 15 October 2023).

Vantage Market Research. 2023. Titanium Dioxide Powder Market – Global Industry Assessment & Forecast. Available at: https://www.vantagemarketresearch. com/industry-report/titanium-dioxide-powder-market-1957. (Accessed 19 October 2023).

Van Arkel, A.E. and J.H. de Boer. 1925. Preparation of pure titanium, zirconium, hafnium, and thorium metal. Zeitschrift für anorganische und allgemeine Chemie. 148: 345–50.

Villarrubia-Gómez, P., S.E. Cornell and J. Fabres. 2018. Marine plastic pollution as a planetary boundary threat – The drifting piece in the sustainability puzzle. Mar. Policy. 96: 213–220.

Wang, H., Y. Dong, M. Zhu, X. Li, A.A. Keller, T. Wang and F. Li. 2015. Heteroaggregation of engineered nanoparticles and kaolin clays in aqueous environments. Water Res. 80: 130–138.

Wang, L.-M., K. Jia, Z.-F. Li, H.-Y. Qi, D.-X. Liu, Y.-J. Liang, S.-L. Hao, F.-Q. Tan and W.-X. Yang. 2023. TiO$_2$ nanoparticles affect spermatogenesis and adhesion junctions via the ROS-mediated mTOR signalling pathway in *Eriocheir sinensis* testes. Environ. Pollut. 331: 121952.

Wang, P., L. Zhao, Y. Huang, W. Qian, X. Zhu, Z. Wang and Z. Cai. 2021. Combined toxicity of nano-TiO$_2$ and Cd^{2+} to *Scenedesmus obliquus*: Effects at different concentration ratios. J. Hazard. Mater. 418: 126354.

Wang, Z., B. Xia, B. Chen, X. Sun, L. Zhu, J. Zhao, P. Du and B. Xing. 2017. Trophic transfer of TiO$_2$ nanoparticles from marine microalga (*Nitzschia closterium*) to scallop (*Chlamys farreri*) and related toxicity. Environm. Sci. Nano. 4: 415–424.

Wang, Z., L. Yin, J. Zhao and B. Xing. 2016. Trophic transfer and accumulation of TiO$_2$ nanoparticles from clamworm (*Perinereis aibuhitensis*) to juvenile turbot (*Scophthalmus maximus*) along a marine benthic food chain. Water Res. 95: 250–259.

Watari, T., K. Nansai K. and Nakajima. 2020. Review of critical metal dynamics to 2050 for 48 elements. Resour. Conserv. Recy. 155: 104669.

Weeks, M.E. 1932. The discovery of the elements. XI. Some elements isolated with the aid of potassium and sodium: Zirconium, titanium, cerium, and thorium. J. Chem. Educ. 9: 1231.

Weir, A., P. Westerhoff, L. Fabricius, K. Hristovski and N. von Goetz. 2012. Titanium dioxide nanoparticles in food and personal care products. Environ. Sci. Technol. 46: 2242–2250.

Wise, J., W. Thompson, S. Wise, C. LaCerte, J. Wise, C. Gianios, C. Perkins, T. Zheng, L. Benedict, M.D. Mason, R. Payne and I. Kerr. 2011. A global assessment of gold, titanium, strontium and barium pollution using sperm whales (*Physeter Macrocephalus*) as an Indicator species. Journal of Ecosystem and Ecography. 1:1.

Wold, A. 1993. Photocatalytic properties of titanium dioxide (TiO$_2$). Chem. Mater. 5: 3.

Wooldridge, S.A. 2010. Is the coral-algae symbiosis really 'mutually beneficial' for the partners? BioEssays. 32: 615–625.

Wu, F., M. Seib, S. Mauel, S. Klinzing and A.L. Hicks. 2020. A citizen science approach estimating titanium dioxide released from personal care products. PLOS ONE. 15: e0235988.

Wu, X. and J. Zhang. 2006. Geographical distribution and characteristics of titanium resources in China. Titanium Ind. Prog. 23: 8.

Xiao, B., Y. Zhang, X. Wang, M. Chen, B. Sun, T. Zhang and L. Zhu. 2019. Occurrence and trophic transfer of nanoparticulate Ag and Ti in the natural aquatic food web of Taihu Lake, China. Environm. Sci.: Nano. 6: 3431–3441.

Yang, W.-W., Y. Wang, B. Huang, N.-X. Wang, Z.-B. Wei, J. Luo, A.-J. Miao and L.-Y. Yang. 2014. TiO_2 nanoparticles act as a carrier of Cd bioaccumulation in the ciliate *Tetrahymena thermophila*. Environ. Sci. Technol. 48: 7568–7575.

Yu, Z., R. Hao, L. Zhang and Y. Zhu. 2018. Effects of TiO_2, SiO_2, Ag and CdTe/CdS quantum dots nanoparticles on toxicity of cadmium towards *Chlamydomonas reinhardtii*. Ecotox. Environ. Safe. 156: 75–86.

Zainy, F.M.A. 2017. Heavy metals in lipstick products marketed in Saudi Arabia. J. Cosmet. Dermatol. Sci. Appl. 7: 336–348.

Zhang, H., Y. Chen, J. Wang, Y. Wang, L. Wang and Z. Duan. 2022. Effects of temperature on the toxicity of waterborne nanoparticles under global warming: Facts and mechanisms. Mar. Environ. Res. 181: 105757.

Zhang, J., M. Wages, S.B. Cox, J.D. Maul, Y. Li, M. Barnes, L. Hope-Weeks and G.P. Cobb. 2012. Effect of titanium dioxide nanomaterials and ultraviolet light coexposure on African clawed frogs (*Xenopus laevis*). Environ. Toxicol. Chem. 31: 176–183.

Zhang, S., R. Deng, D. Lin and F. Wu. 2017. Distinct toxic interactions of TiO_2 nanoparticles with four coexisting organochlorine contaminants on algae. Nanotoxicology. 11: 1115–1126.

Zhang, X., H. Sun, Z. Zhang, Q. Niu, Y. Chen and J.C. Crittenden. 2007. Enhanced bioaccumulation of cadmium in carp in the presence of titanium dioxide nanoparticles. Chemosphere. 67: 160–166.

Zhang, Y., Y.R. Leu, R.J. Aitken and M. Riediker. 2015. Inventory of engineered nanoparticle-containing consumer products available in the Singapore retail market and likelihood of release into the aquatic environment. Int. J. Env. Res. Pub. He. 12: 8717–8743.

Zhang, Y. and G.G. Goss. 2021. The "Trojan Horse" effect of nanoplastics: potentiation of polycyclic aromatic hydrocarbon uptake in rainbow trout and the mitigating effects of natural organic matter. Environm. Sci.: Nano. 8: 3685–3698.

Zhu, X., J. Wang, X. Zhang, Y. Chang and Y. Chen. 2010. Trophic transfer of TiO_2 nanoparticles from daphnia to zebrafish in a simplified freshwater food chain. Chemosphere. 79: 928–933.

Zhu, X., J. Zhou and Z. Cai. 2011. TiO_2 Nanoparticles in the marine environment: Impact on the toxicity of tributyltin to abalone (*Haliotis diversicolor supertexta*) embryos. Environ. Sci. Technol. 45: 3753–3758.

Index

For Product Safety Concerns and Information please contact our EU
representative GPSR@taylorandfrancis.com
Taylor & Francis Verlag GmbH, Kaufingerstraße 24, 80331 München, Germany

www.ingramcontent.com/pod-product-compliance
Lightning Source LLC
Chambersburg PA
CBHW061609220326
41598CB00024BC/3512